MINECRAFT

MATHS
OFFICIAL WORKBOOK
AGES 8-9

**DAN LIPSCOMBE
AND LEISA BOVEY**

INTRODUCTION

HOW TO USE THIS BOOK

Welcome to an exciting educational experience! Your child will go on a series of adventures through the amazing world of Minecraft, improving their maths skills along the way. Matched to the National Curriculum for maths for ages 8–9 (Year 4), this workbook takes your child into fascinating landscapes where our heroes Maya and Oscar embark on building projects and daring treasure hunts…all while keeping those pesky mobs at bay!

As each adventure unfolds, your child will complete topic-based questions worth a certain number of emeralds . These can then be 'traded in' on the final page. The more challenging questions are marked with this icon  to stretch your child's learning. Answers are included at the back of the book.

MEET OUR HEROES

Maya can be quite shy. She likes to spend time in her own company, often with a good book. She enjoys building and discovering new materials. Her favourite block is quartz. Don't let her quiet personality fool you, though. Maya is very handy with a bow and arrow and fights never scare her. She is smart and will retreat to home if she knows the fight can't be won.

Oscar is very athletic. Not many people can keep up with him when he is exploring. He never seems to stand still! He is forever running, jumping and swimming – eager to reach something new and exciting. Sometimes he is so active that he forgets to eat. Oscar quite likes Endermen because they are also fast but, if he has to fight one, you can count on him!

First published in 2021 by Collins
An imprint of HarperCollins*Publishers*
1 London Bridge Street, London, SE1 9GF

HarperCollins*Publishers*
Macken House, 39/40 Mayor Street Upper,
Dublin 1, D01 C9W8, Ireland

Publisher: Fiona McGlade
Authors: Dan Lipscombe and Leisa Bovey
Project management: Richard Toms
Design: Ian Wrigley and Sarah Duxbury
Typesetting: Nicola Lancashire at Rose and Thorn
Creative Services

Special thanks to Alex Wiltshire, Sherin Kwan and Marie-Louise Bengtsson at Mojang and the team at Farshore

Production: Karen Nulty

ISBN: 978-0-00-846277-2
British Library Cataloguing in Publication Data.
A CIP record of this book is available from the British Library.
10 9 8 7 6 5 4
Printed in the United Kingdom

MIX
Paper | Supporting responsible forestry
FSC www.fsc.org FSC™ C007454

This book contains FSC™ certified paper and other controlled sources to ensure responsible forest management.

For more information visit: www.harpercollins.co.uk/green

CONTENTS

NUMBER AND PLACE VALUE

MYSTERIOUS ILLAGERS

Lots of mobs spawn among the shadows of the dark forest, even by day. In this particular forest, the roof of a large mansion towers above the trees. There are many rooms inside. Anyone passing by will hear odd noises and shouts from the illagers who live there. What happens inside these mansions, nobody really knows.

MAYA'S ON THEIR TRAIL

Rumours suggest that the illagers were holding magical treasures as they entered the nearby dark forest. Being the heroine that she is, Maya decides to investigate. Very little is known about the illagers but Maya has been led to believe that the largest door on the upper floor of their mansion will reveal much more.

COUNTING IN MULTIPLES

Maya sets off to find the large mansion deep in the dark forest. As she steps into the forest, she counts the trees in multiples to get a better understanding of her exact position.

1

Write the missing numbers in these sequences.

a) 50   150   250   ☐   450   ☐

b) 0   25   50   ☐   100   ☐

c) 0   9   18   ☐   36   ☐

d) 0   6   ☐   ☐   24   30

2

What is the rule for each sequence of numbers?

a) 14   21   28   35   42   **Rule:** ....

b) 60   66   72   78   84   **Rule:** ....

c) 500   475   450   425   400   **Rule:** ....

d) 108   99   90   81   72   **Rule:** ....

3

Here is a number grid. The numbers increase by 6 as you move right and increase by 7 as you move down. Fill in the missing numbers.

Add 6 →

Add 7 ↓

24	30	36	42	48
31		43	49	55
	44	50		62
45	51		63	69

PLACE VALUE

As Maya moves through the trees, she stops now and then to harvest some wood – just in case she needs to make extra tools. Because the trees are tightly packed, she bumps into zombies here and there.

1

Write the digit that is in the thousands place in each number.

a) 6,138

b) 4,170

c) 3,106

2

Write in words the value of the digit 7 in each number.

a) 3,217 ....

b) 7,408 ....

c) 8,713 ....

d) 9,071 ....

3

Here are four signs with numbers on them:

  2

What is the largest and the smallest four-digit number you can make using the signs?

Largest number:

Smallest number:

Maya has never seen so many trees! Not only do they grow close together, but some are also incredibly tall. To her surprise, she even finds some towering mushrooms, which she harvests. She loves mushroom stew!

 4

Write the symbol **<**, **>** or **=** in each box to make the statements true.

a) 9,391 ☐ 9,084

b) 3,907 ☐ 3,711

c) 8,798 ☐ 8,887

d) 3,010 ☐ 3,515

 5

The numbers of several forest objects are listed in this table.

Write the numbers in order from smallest to largest.

Item		Number
Brown mushrooms		8,023
Dark oak leaves		13,667
Rose bushes		11,569
Red mushrooms		8,467
Dark oak logs		17,421

 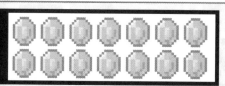

WORKING WITH NUMBERS BEYOND 1,000

Further ahead, Maya can see the trees begin to thin out. As she gets closer, she finds herself looking down into a ravine. It's a long way down. At the bottom she can see enormous lakes of lava. In some places, obsidian has formed where water flows onto lava.

1

Fill in the missing numbers in each sequence.

a)   1,250   2,250   [ ]   4,250   [ ]

b)   3,465   [ ]   5,465   6,465   [ ]

c)   9,852   8,852   [ ]   6,852   [ ]

d)   7,648   6,648   [ ]   [ ]   3,648

Maya carefully builds a long wooden bridge across the ravine, making sure she doesn't fall off.

2

Find 1,000 less than each number.

a)   6,237   [ ]   b)   1,335   [ ]

c)   4,012   [ ]   d)   5,066   [ ]

3

Find 1,000 more than each number.

a)   4,758   [ ]   b)   8,301   [ ]

c)   3,437   [ ]   d)   8,576   [ ]

Safely across to the other side of the ravine, Maya breathes a sigh of relief. Above the trees, she can now see the roof of the mansion. After sneaking closer, Maya peeks through a window and sees a chest. She wonders how much loot might be inside.

4

Complete the table to show 1,000 less and 1,000 more of each starting number.

1,000 less	Starting number	1,000 more
4,950	5,950	6,950
	3,264	
	2,621	
7,573		
		9,482

5

 One of the illagers has four chests of gold ingots.

 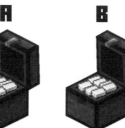 

A   B   C   D

5,537   3,687   7,494   4,872

The illager takes 1,000 gold ingots from chest A and puts them in chest B.

They then take 100 gold ingots from chest B and put them in chest C.

They then take 1,000 gold ingots from chest C and put them in chest D.

They then take out 10 gold ingots from chest D.

a) How many gold ingots does each chest have now?

A: _____   B: _____   C: _____   D: _____

b) Order the new quantities in each chest from largest to smallest.

....

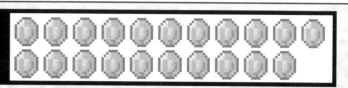

REPRESENTING NUMBERS

You can represent numbers in different ways. Example: 1,124 = 1,000 + 100 + 20 + 4

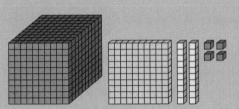

Th	H	T	O

Dienes blocks | **Place value counters** | **Part-whole model**

Maya walks around the mansion to look for a safe way to enter. Looking through the windows, each room seems to have lots of items.

1

a) Write down the number that is shown by these dienes blocks.

b) Show how you can use place value counters to represent the number 2,435.

Th	H	T	O

c) Show the number 7,697 using a part-whole model.

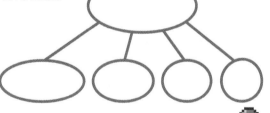

d) 4,579 =

☐ + ☐ + ☐ + ☐

2

Write these numbers in the correct place on the number line.

       2,615   **2,016**   **3,591**   **1,203**

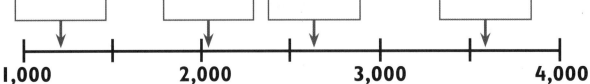

1,000   2,000   3,000   4,000

Maya finds an unguarded door and enters a corridor. She dashes up the stairs ahead of her. She can hear chanting and strange noises. There are illagers moving on the floor below and they sound angry. She needs to hide in one of the rooms, but which one? Maya remembers what she heard about the largest door on the upper floor.

3

Write the number on each door in words.

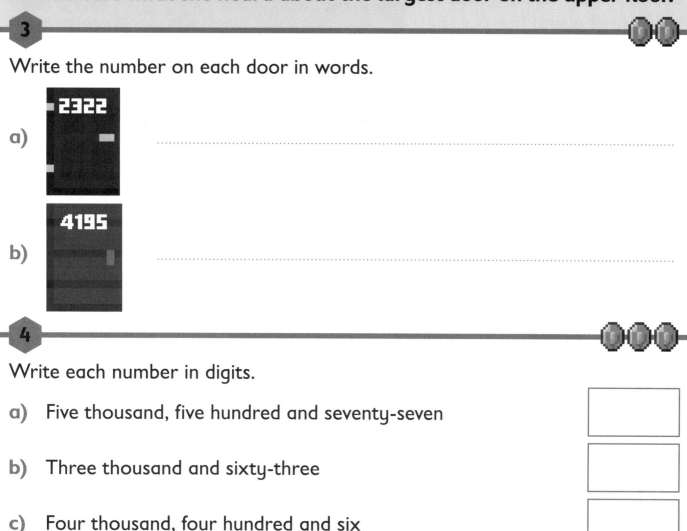

a) 2322 ....

b) 4195 ....

4

Write each number in digits.

a) Five thousand, five hundred and seventy-seven

b) Three thousand and sixty-three

c) Four thousand, four hundred and six

5

Write down an estimate for each number shown on the number line.

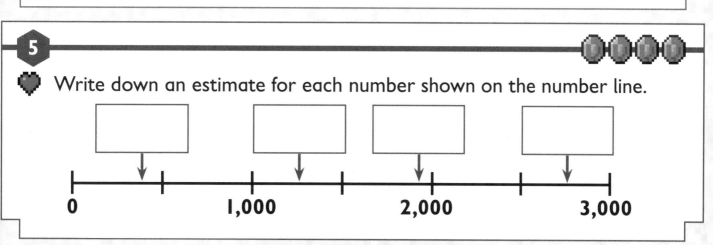

0   1,000   2,000   3,000

COLOUR IN HOW MANY
EMERALDS YOU EARNED

ROUNDING NUMBERS

Maya throws open the door and darts inside. She shuts the door behind her but then sees an illager in the room with four lecterns, which have books laid open. The illager is holding a golden totem.

I

Here are four numbers:

a) Write each number in the correct place on the number line.

200 210 220 230 240 250 260 270 280 290 300

b) Round each number to the nearest 10 and to the nearest 100.

Number	Nearest 10	Nearest 100
242		
288		
214		
209		

The illager has not noticed Maya. They are too busy mumbling about numbers and pacing back and forth. Then Maya notices symbols floating through the air... floating towards the illager! She hadn't spotted the enchantment table. Maya thinks the illager might be trying to somehow enchant the totem. She doesn't like the look of this. She rushes over to destroy the lecterns and enchantment table.

2

Here are four numbers: **6,384   6,828   6,782   6,021**

a) Write each number in the correct place on the number line.

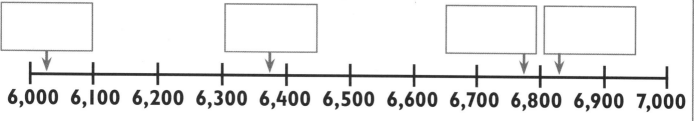

6,000 6,100 6,200 6,300 6,400 6,500 6,600 6,700 6,800 6,900 7,000

b) Round each number to the nearest 10, 100 and 1,000.

Number	Nearest 10	Nearest 100	Nearest 1,000
6,384			
6,828			
6,782			
6,021			

3

Here are four digits swirling around an enchantment table:

4   7

5   3

Use the digits to make four different 4-digit numbers that will all round to 5,000 to the nearest 1,000.

NEGATIVE NUMBERS

With the lecterns and enchantment table broken, the illager snaps out of their trance and turns on Maya. The illager raises their arms. Purple sparks fly from their fingers as large spikes shoot from the floor. Maya barely gets out of the way in time. This is not a normal illager... it's an evoker!

1

Write the numbers marked by the arrows on the number line.

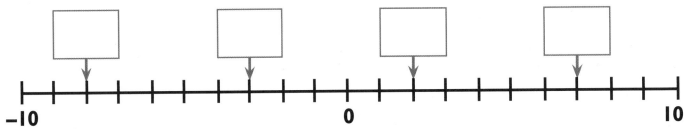

−10   0   10

2

Write down the final number you reach in each of these calculations.
Use the number line above to help you.

a) Start at 4, count back 7.

b) Start at 8, count back 10.

c) Start at 3, count back 4.

d) Start at −2, count back 2.

3

Calculate:

a) $2 - 9 =$

b) $6 - 8 =$

c) $3 - 7 =$

The evoker is conjuring vexes to fight for them. Maya is being hit from each side and is hurt. She quickly drinks a Potion of Invisibility and vanishes. The evoker is confused but raises their hands once more. The room begins to grow cold. It's freezing!

4

Fill in the missing numbers on the thermometer.

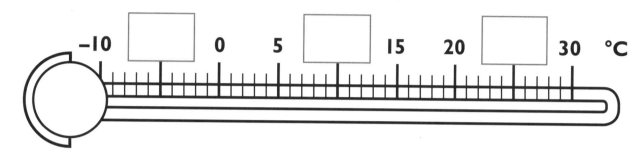

-10   0   5   15   20   30   °C

5

Find the temperature after these changes. Use the thermometer in question 4 to help you.

a) The temperature is 3°C and falls by 8°C   °C

b) The temperature is 11°C and falls by 13°C   °C

c) The temperature is 15°C and falls by 23°C   °C

d) The temperature is 5°C and falls by 10°C   °C

6

 Maya is waiting for the right moment to attack. That will be when the evoker is charging their spells. Maya counts down in 7s.

The first number in her sequence is 14.

What is the sixth number in her sequence?

ROMAN NUMERALS

| Remember: | I = 1 | V = 5 | X = 10 | L = 50 | C = 100 |

Maya gives it all she has got left. Drinking a Potion of Healing, she also throws a splash Potion of Poison at the evoker. Maya ignores the vexes and attacks the evoker with her sword. Still confused, the evoker can't see Maya and has dropped the totem in the struggle! One more potion will help. Maya swallows a Potion of Strength and inflicts more damage, defeating the evoker.

1

Write these Roman numerals in digits.

a) LVIII ☐

b) XC ☐

c) XXIII ☐

d) LXXIX ☐

2

Write these numbers in Roman numerals.

a) 31 ....

b) 95 ....

c) 68 ....

d) 15 ....

3

♥ Written on the back of the totem are Roman numerals.
Help Maya solve the problems below and write the answers in digits.

a) C + XLII + XLV = ☐

b) C − LXXXIX = ☐

c) C + XCVI + LIV = ☐

COLOUR IN HOW MANY EMERALDS YOU EARNED

ADVENTURE ROUND-UP

MANSION FALLS QUIET

Everything is silent in the mansion now that the evoker has gone. Maya holds the totem in her hands. After decoding the Roman numerals, she realises they are co-ordinates. Perhaps they lead to treasure? Or maybe to where the totem was originally discovered by the evoker?

POTION POWER

Maya hadn't expected such a tough fight but her potion-making skills saved her. As she walks through the mansion towards the exit, she sees items all over the floor. It's as if the illagers have just disappeared into thin air.

TOTEM OF UNDYING

Maya leaves the dark forest with it returned to peace. Back in the village, she shows the totem to her friend, Oscar. He tells her it's a rare treasure — the totem of undying. It's said that whoever holds the totem will be saved from death, but only once, as the totem breaks after it has been used. Maya tucks it away for safe keeping in her inventory.

ADDITION, SUBTRACTION, MULTIPLICATION AND DIVISION

DESTINATION DESERT

Oscar has built a home in the mountains of the badlands, which lie next to a vast desert. Tired of mining only gold from the ground beneath him, he rides out into the desert sands on the back of his mule. With plenty of food packed into the mule's saddle chests, he finds a cave and sets up a small camp.

HUSKS ON THE HUNT

The deserts of the Overworld are lonely places. If a figure can be seen in the distance, it's likely to be a husk looking to munch any lost heroes. The husk is a strange creature; a zombie that has grown used to the burning sunshine and so can hunt in the day.

ABANDONED PYRAMIDS

Long abandoned desert pyramids sit quiet. A lucky hero who stumbles upon these large buildings may find treasure hidden inside…if they can avoid the traps.

SAND TRAPS

Anyone venturing into the desert must bring plenty of food. Nothing grows except for cactus plants and brown bushes that are now dying. It's quite easy to see why so many people avoid setting up home in the desert. Even digging a hole here can be dangerous as the sand caves in, leaving the explorer without air.

ADDITION AND SUBTRACTION PRACTICE

Oscar builds a small shelter with a storage chest and campfire. In the entrance of the cave, he can see layers of materials which make up the desert floor: sand, sandstone and smooth stone. He is interested in harvesting some sandstone as he wants to build a mini temple at home.

1

Work out these addition calculations.

a)
```
   8 2 2
 + 1 7 6
 _____
```

b)
```
   5 3 8
 + 2 5 5
 _____
```

c)
```
   6 4 1 5
 +   2 2 4
 _____
```

d)
```
   2 6 4 4
 + 4 9 3 2
 _____
```

e)
```
   5 3 4 3
 + 2 3 5 9
 _____
```

f)
```
   8 3 2 7
 + 1 4 3 5
 _____
```

2

Work out these subtraction calculations.

a)
```
   4 1 3
 - 2 0 3
 _____
```

b)
```
   3 3 2
 - 1 0 4
 _____
```

c)
```
   3 6 1 6
 -   4 0 4
 _____
```

d)
```
   5 5 0 8
 - 3 2 2 6
 _____
```

e)
```
   2 7 2 5
 - 1 8 0 3
 _____
```

f)
```
   9 7 1 1
 - 7 6 0 7
 _____
```

ESTIMATING AND CHECKING ADDITION AND SUBTRACTION

Digging away the sandstone, Oscar has to be careful of the sand above caving in. If he were to get stuck under the sand, he could lose air and faint. He always leaves a one-block layer of sandstone to hold up the blocks of sand. So far, he has harvested quite a lot of materials.

1

The pictures on the left show how many sandstone, coal, cobblestone and iron ore Oscar has mined.

Draw lines to join each picture with the best estimate of the number of items.

Sandstone:

0   100   200

Coal:

0   50   100

Cobblestone:

0   100   200

Iron ore:

0   25   50

175 items

125 items

10 items

30 items

2

Draw lines to join the calculations with the best estimate.

Calculations

3,512 + 4,894	2,967 + 7,370	3,459 + 3,202	2,732 + 7,998

3,000 + 7,400	3,500 + 3,200	2,700 + 8,000	3,500 + 5,000

Estimates

When mining, Oscar digs away hundreds of blocks to reach a depth at which he might find diamonds.

3

Round each number to the nearest 100 to work out an estimate for the answer. Then work out the actual answer. Check your answer by writing the inverse calculation.

a)
```
   3 0 2 2
 + 2 5 3 6
 _____

 _____
```

Estimate:

.... + ... = ...

Inverse:

b)
```
   8 6 0 9
 - 4 4 7 3
 _____

 _____
```

Estimate:

.... + ... = ...

Inverse:

4

 Oscar can trade unwanted stone with a mason villager. Oscar has 1,000 granite blocks.

Shown right are prices for one stack each of books, cooked porkchops and golden carrots.

One stack of of books: 318 granite

One stack of cooked porkchops: 257 granite

One stack of golden carrots: 468 granite

Estimate if he has enough granite to trade for the following.

a) Three stacks of books ....

b) One stack of books and two stacks of cooked porkchops

....

c) One stack of golden carrots and two stacks of books

....

COLOUR IN HOW MANY EMERALDS YOU EARNED

ADDITION AND SUBTRACTION PROBLEMS

Oscar returns to the surface and decides to explore the surrounding area. His campfire will show him where to return to, so he doesn't get lost. After a while, Oscar discovers a desert pyramid.

 1

Complete the missing numbers on the pyramid. Each missing number is the total of the numbers in the two blocks below it.

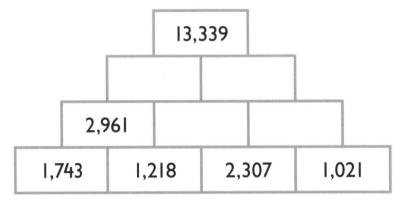

13,339

2,961

1,743   1,218   2,307   1,021

2

After exploring the pyramid and mining around it, Oscar has more valuable materials.

a) Oscar had 4,459 sand. He then mined 1,426 more.

How many sand does he have now? ☐ sand

b) He had 6,273 sandstone. He then mined 2,495 more.

How many sandstone does he have now? ☐ sandstone

3

Sometime later, to save space in the chests being carried by his mule, Oscar throws some materials into a lava pool to destroy them.

a) Oscar had 5,672 sand. He then burned 3,712 of them.

How many sand does he have left? ☐ sand

b) He had 8,756 dirt. He then burned 2,328 of them.

How many dirt does he have left? ☐ dirt

Oscar has been in the desert for a few days but still hasn't seen anyone. He is busy digging away another layer of stone in the cave when a voice calls down. He exits the cave with his sword drawn, but there is no danger. It's a wandering trader.

4

a) On Monday, the trader had 1,587 green dye. On Tuesday, they traded away 372 of them. On Wednesday they obtained 1,283 more.

How many green dye does the trader have now? ⬚ green dye

b) The trader started with 7,267 kelp. They then got 1,624 more. They later traded away 2,413.

How many kelp does the trader have now? ⬚ kelp

c) The trader had 8,976 slimeballs two weeks ago. Last week they traded away 5,458 of them. This week they traded away another 1,812 slimeballs.

How many slimeballs does the trader have now? ⬚ slimeballs

5

 Fill in the missing numbers in these calculations.

a)
```
   3 _ 7 _
 + _ 2 _ 0
 ---------
   7 1 8 2
```

b)
```
   _ 0 5 _
 + 3 _ 0 3
 ---------
   7 6 6 1
```

c)
```
   4 _ 9 4
 - _ 8 1 2
 ---------
   2 8 8 2
```

d)
```
   4 _ _ 6
 - _ 4 5 7
 ---------
   3 1 2 9
```

COLOUR IN HOW MANY EMERALDS YOU EARNED

MULTIPLICATION AND DIVISION FACTS

Oscar notices the trader has plenty of slimeballs in stock. He trades 8 emeralds for 2 slimeballs. These slimeballs can be used to craft sticky pistons or leads for guiding animals. The wandering trader is now ready to move on. Help the trader to do some calculations with his stash of emeralds.

1

Fill in the missing numbers in each calculation.

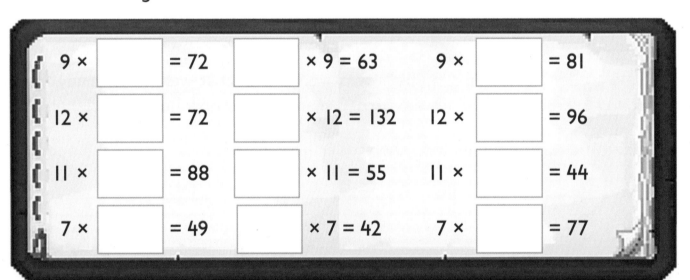

9 × ☐ = 72   ☐ × 9 = 63   9 × ☐ = 81

12 × ☐ = 72   ☐ × 12 = 132   12 × ☐ = 96

11 × ☐ = 88   ☐ × 11 = 55   11 × ☐ = 44

7 × ☐ = 49   ☐ × 7 = 42   7 × ☐ = 77

2

Find the answer to each calculation.

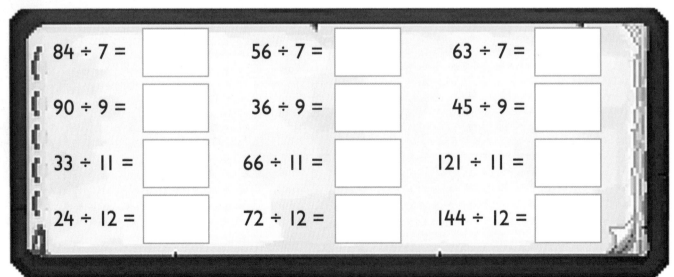

84 ÷ 7 =   56 ÷ 7 =   63 ÷ 7 =

90 ÷ 9 =   36 ÷ 9 =   45 ÷ 9 =

33 ÷ 11 =   66 ÷ 11 =   121 ÷ 11 =

24 ÷ 12 =   72 ÷ 12 =   144 ÷ 12 =

Oscar waves goodbye to the trader, walks back to the bottom of his mine and starts digging out rocks. He hopes to see a glimpse of gems or metals.

3

Here is some iron ore. Fill in the spaces in each sentence and calculation.

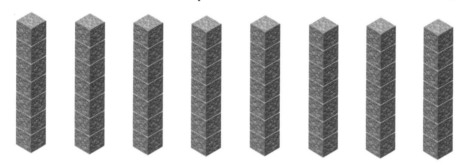

a) There are _____ iron ore in total.

b) There are _____ groups of 7.

_____ × 7 = _____

_____ ÷ 7 = _____

c) There are _____ groups of 8.

_____ × 8 = _____

_____ ÷ 8 = _____

4

There are 9 stacks of coal. Each stack has 7 blocks.

Complete the fact family.

a) _____ × _____ = _____   b) _____ ÷ _____ = _____

_____ × _____ = _____   _____ ÷ _____ = _____

25

MENTAL MULTIPLICATION AND DIVISION

Just when **Oscar** thinks he might never find diamonds, he breaks away stone to reveal a light blue sparkle. Finally, here they are! Oscar keeps digging and finding more diamonds, along with some emeralds!

1

Find the answer to each calculation.

a) $7 \times 1 =$ ▢ $0 \times 12 =$ ▢ $78 \times 1 =$ ▢

b) $30 \times 3 =$ ▢ $70 \times 5 =$ ▢ $40 \times 0 =$ ▢

2

Fill in the missing numbers.

a) $160 \div$ ▢ $= 40$ $300 \div$ ▢ $= 50$

b) $490 \div$ ▢ $= 70$ $210 \div$ ▢ $= 70$

3

Find the answer to each calculation.

a) $1 \times 9 \times 4 =$ ▢   b) $5 \times 2 \times 7 =$ ▢

c) $6 \times 5 \times 4 =$ ▢   d) $11 \times 9 \times 2 =$ ▢

Oscar has done enough mining. He loads up the mule's chests with diamonds, emeralds, iron and gold. It's unlikely that he will return to this part of the desert, so he will leave a small building – a castle-shaped shelter – for future explorers to use. Answer the questions below to help him work out the number of different blocks he needs to build it.

4

Complete the calculations using the given information.

*Example: Given 2 × 4 = 8, then 800 ÷ 4 = **200***

a) Given 5 × 11 = 55, then 550 ÷ 110 = ☐

b) Given 12 ÷ 3 = 4, then 30 × 4 = ☐

5

 Show how you can use times tables facts you know to solve each question mentally.

*Example: 15 × 5 = **(10 × 5)** + **(5 × 5)** = **50 + 25 = 75***

a) 14 × 8 = (☐ × ☐) + (☐ × ☐)

= ☐ + ☐ = ☐

b) 16 × 5 = (☐ × ☐) + (☐ × ☐)

= ☐ + ☐ = ☐

6

Fill in the missing numbers.

a) 3 × 8 × ☐ = 48

b) 4 × ☐ × 3 = 60

c) 7 × ☐ × 2 = 42

d) ☐ × 4 × 8 = 64

FACTORS

Factors are numbers that divide exactly into another number so that they do not leave a remainder. For example, the factors of 6 are: 1, 2, 3, 6.

In building the castle-style shelter, Oscar tries to use as many leftover materials as possible. He doesn't want anything to go to waste.

1

Circle the two correct statements.

9 is a factor of 84

14 is a factor of 28

8 is a factor of 36

25 is a factor of 80

7 is a factor of 56

2

Use diamonds to show the factor pairs of the given number.

Example: Factor pairs of 6.

$6 \times 1 = 6$

$2 \times 3 = 6$

The two factor pairs of 6 are: 1 and 6, and 2 and 3. 6 has four factors: 1, 2, 3, and 6.

Draw the factor pairs of 30. One has been done for you. Then complete the statement.

30 has _____ factors: _____

Oscar's little castle is starting to take shape.

3

Circle the numbers on Oscar's castle that are factors of each given number.

a) Factors of 16

b) Factors of 34

4

Write all the factors of each of these numbers.

a) 15 ....

b) 24 ....

c) 20 ....

5

Complete the Venn diagram to show the factors of 12 and the factors of 18.

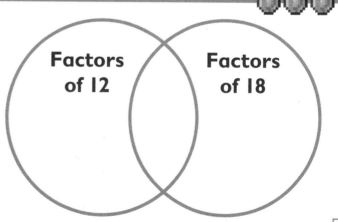

MULTIPLICATION PRACTICE

Oscar adds the finishing touches to his castle. He wants it to be a safe place in the desert, where explorers can stop and rest. In the central room, Oscar places some chests which he fills with food and other supplies for tired adventurers who may pass by.

Find the answer to each multiplication.

a)
```
    3 9
×     5
_____
```

b)
```
    1 7
×     4
_____
```

c)
```
    5 7
×     9
_____
```

d)
```
    6 5
×     5
_____
```

2

Write out these multiplications using the column method and find the answer.

a) 47 × 4

b) 36 × 6

c) 27 × 8

d) 38 × 5

e) 41 × 7

f) 85 × 3

The final thing Oscar does before leaving the desert is write a sign for anyone visiting the castle. He writes: "Rest here. Heal and eat!"

Oscar breaks down his camp, climbs onto his mule and heads for home. Work out how many blocks he has to travel by doing these calculations.

3

Find the answer to each multiplication.

a)
```
  2 1 1
×     8
_____
_____
```

b)
```
  4 2 4
×     3
_____
_____
```

c)
```
  2 3 1
×     6
_____
_____
```

d)
```
  3 0 7
×     4
_____
_____
```

4

Write out these multiplications using the column method and find the answer.

a) 414 × 3

b) 221 × 7

c) 324 × 6

d) 216 × 5

5

 Here are three blocks with digits on them:

  4 7 5

Complete the multiplication calculation to show the largest total you can make with these digits. One digit is already given.

```
  □ □ 3
×     □
_____

_____
```

MULTIPLICATION PROBLEMS

With the desert behind him, Oscar can finally see the lights of his house. On arrival, he unloads his mule and begins putting everything away.

1

There are 27 iron blocks in a large chest. Each large chest costs 40 emeralds.

a) How many iron blocks are in 6 large chests? [ ] iron blocks

b) How many emeralds are needed to buy 6 large chests? [ ] emeralds

c) How many iron blocks are in 9 large chests? [ ] iron blocks

d) How many emeralds are needed to buy 9 large chests? [ ] emeralds

2

A small chest holds 7 diamonds. Each small chest costs 30 emeralds.

a) One of Oscar's neighbours has 77 diamonds.
How many full chests of diamonds does his neighbour have? [ ] chests

b) How many emeralds are needed to buy enough small chests for 77 diamonds? [ ] emeralds

c) Another neighbour would like to have 560 diamonds.
How many small chests does that neighbour need? [ ] chests

d) How many emeralds are needed to buy enough small chests for 560 diamonds? [ ] emeralds

3

a) The distance from Oscar's house to the stable is 28 blocks.
Each block is 1 metre long. How many metres will Oscar travel if he makes three trips there and back? [ ] m

b) The distance from Oscar's house to a lake is 539 metres. The distance from Oscar's house to Maya's house is four times as far.
What is the distance from Oscar's house to Maya's house? [ ] m

COLOUR IN HOW MANY EMERALDS YOU EARNED

ADVENTURE ROUND-UP

RICH REWARDS

It's been a very busy few days in the desert. Oscar found plenty of iron and some gems too. Oscar wanted to build a small, castle-style shelter alongside the cave that he mined. He hopes other explorers find it useful in future.

WELL-EARNED REST

As much as Oscar loves mining and building, it feels good to be home. His mule has happily trotted off to the warmth of its stable. It's nice to unpack everything, treat himself to a slice of cake and sleep well in his own bed.

SUNFLOWERS STAND TALL

This plains biome is bursting with flowers. Sunflowers, to be exact. An adventurer cannot wander too far without seeing these tall, cheery plants. The sunflowers face east, where the sun rises. Bees buzz from sunflower to sunflower. They are collecting pollen to take back to their hive, where they will make yummy honey. This honey can be bottled by anyone passing by.

PERFECT PLAINS

Many adventurers consider the plains to be the best place to set up a permanent home. There is plenty of flat land for building and for farming. Fish can be caught in the rivers and natural caves offer mining opportunities.

BIG PLANS

Maya already has a small house in the sunflower plains but she needs more space. She therefore plans to build something bigger, with extra rooms, a garden, more space for animals to live and a separate building for potion making and enchanting items.

EQUIVALENT FRACTIONS

$\frac{1}{2}$ and $\frac{2}{4}$ are examples of equivalent fractions. They have the same value.

Maya already has plenty of building materials from her travels through other biomes. The first step in building the new house is to place the foundation layer. Answer these questions to help Maya work out what fractions she will use of different materials she has available.

1

Shade in the bottom diagram in each pair so that it shows an equivalent fraction to the diagram above.

a)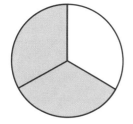

b)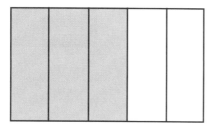

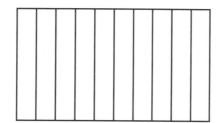

c)

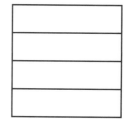

2

Fill in the boxes to make each statement true.

a) $\frac{1}{3} = \frac{\boxed{}}{9}$

b) $\frac{4}{5} = \frac{\boxed{}}{20}$

c) $\frac{7}{8} = \frac{\boxed{}}{16}$

d) $\frac{3}{4} = \frac{\boxed{}}{16}$

3

 Fill in the boxes to show fractions equivalent to $\frac{1}{5}$

$$\frac{1}{5} = \frac{\boxed{}}{10} = \frac{4}{\boxed{}} = \frac{\boxed{}}{25} = \frac{\boxed{}}{100}$$

COLOUR IN HOW MANY EMERALDS YOU EARNED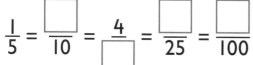

ADDING AND SUBTRACTING FRACTIONS

You can use diagrams to help you add and subtract fractions.

For example:

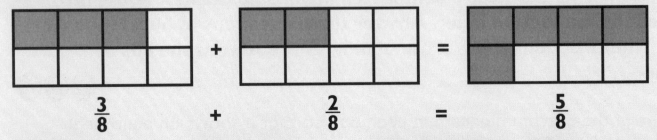

$$\frac{3}{8} \quad + \quad \frac{2}{8} \quad = \quad \frac{5}{8}$$

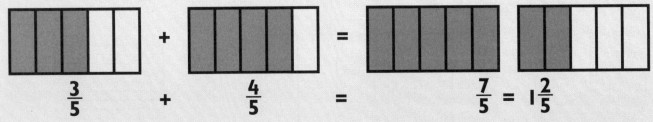

$$\frac{3}{5} \quad + \quad \frac{4}{5} \quad = \quad \frac{7}{5} = 1\frac{2}{5}$$

Once the foundations and floor are laid, Maya begins to work on the walls and ceilings. She wants to use wood for the walls. The ceiling will be stone, but she will use wooden logs to create beams across it.

 1

Shade in the diagram to show how you can add $\frac{2}{6} + \frac{3}{6}$

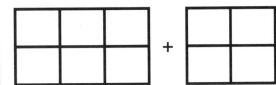 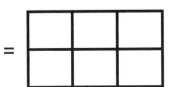

 2

Work out the answers to these fraction additions.

a) $\frac{4}{9} + \frac{2}{9} = \boxed{-}$

b) $\frac{2}{12} + \frac{5}{12} = \boxed{-}$

c) $\frac{1}{6} + \frac{2}{6} = \boxed{-}$

d) $\frac{2}{10} + \frac{7}{10} = \boxed{-}$

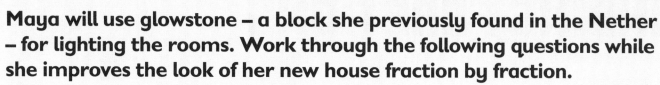

Maya will use glowstone – a block she previously found in the Nether – for lighting the rooms. Work through the following questions while she improves the look of her new house fraction by fraction.

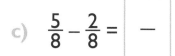

3

Work out the answers to these fraction subtractions.

a) $\frac{5}{7} - \frac{2}{7} =$ ⬚

b) $\frac{4}{9} - \frac{1}{9} =$ ⬚

c) $\frac{5}{8} - \frac{2}{8} =$ ⬚

d) $\frac{5}{7} - \frac{3}{7} =$ ⬚

4

Shade in the diagram to show how you can add $\frac{3}{4} + \frac{2}{4}$

 **+**  **=**

5

Work out the answers to these calculations.

a) $\frac{3}{8} + \frac{6}{8} =$ ⬚

b) $\frac{6}{9} + \frac{8}{9} =$ ⬚

c) $\frac{5}{6} + \frac{4}{6} =$ ⬚

d) $\frac{2}{6} + \frac{5}{6} =$ ⬚

6

♥ Fill in the missing fractions in these calculations.

a) $\frac{3}{7} +$ ⬚ $= \frac{6}{7}$

b) $\frac{4}{12} +$ ⬚ $= \frac{9}{12}$

c) $\frac{5}{8} +$ ⬚ $= \frac{11}{8} = 1\frac{3}{8}$

d) $\frac{10}{13} -$ ⬚ $= \frac{8}{13}$

COLOUR IN HOW MANY EMERALDS YOU EARNED

FINDING FRACTIONS

Maya steps outside to build a fence around the area which will become her garden. Rather than growing flowers, she will use fractions to divide the space for planting different food crops.

1

Here is a bar model representing 64:

a) Split the bar model into eighths.

64

b) Find $\frac{1}{8}$ of 64. ☐   c) Find $\frac{3}{8}$ of 64. ☐   d) Find $\frac{7}{8}$ of 64. ☐

2

Fill in the boxes in these sentences.

a) To find $\frac{1}{4}$ of a number, divide the number by ☐ .

b) $\frac{1}{4}$ of 16 = ☐

c) To find $\frac{3}{4}$ of a number, multiply $\frac{1}{4}$ of the number by ☐ .

d) $\frac{3}{4}$ of 16 = ☐

3

Answer these calculations to help Maya work out how many of each crop she will plant.

a) $\frac{3}{4}$ of 60 potatoes = ☐ potatoes

b) $\frac{11}{12}$ of 48 carrots = ☐ carrots

c) $\frac{17}{20}$ of 40 wheat seeds = ☐ wheat seeds

d) $\frac{4}{9}$ of 81 beetroot seeds = ☐ beetroot seeds

Maya plants the crops and adds some small streams to help keep the soil damp and healthy. While she is outside, she will work on a barn to keep sheep. Maya likes different colours of wool, so she will breed sheep and dye them lots of colours.

4

Maya has plenty of wood for the barn. Tick the statement in each pair that gives a bigger answer.

a) $\frac{3}{4}$ of 20 oak planks ☐ $\frac{3}{8}$ of 24 oak planks ☐

b) $\frac{4}{5}$ of 35 jungle planks ☐ $\frac{3}{5}$ of 40 jungle planks ☐

c) $\frac{2}{3}$ of 60 acacia planks ☐ $\frac{3}{4}$ of 60 acacia planks ☐

d) $\frac{6}{7}$ of 28 spruce planks ☐ $\frac{7}{12}$ of 48 spruce planks ☐

5

This picture shows 7 blocks that form part of a row of planks in the barn. The rest of the row is not shown.

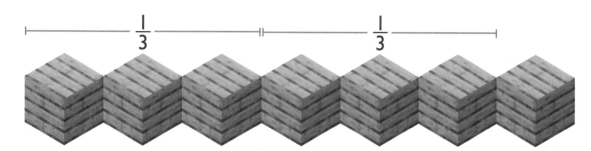

How many blocks make up the full row of planks? ☐ blocks

6

Fill in the boxes to complete the calculations.

a) $\frac{3}{4}$ of 16 = $\frac{1}{3}$ of ☐   b) $\frac{4}{5}$ of 40 = $\frac{1}{2}$ of ☐

HUNDREDTHS

Maya wants to leave plenty of grassy space for the sheep to graze. Without grass, their wool does not grow back once sheared.

1

These grids show three fields that each have 100 blocks. Grass has grown on the blocks shown in green. What fraction of each field is made up of grass?

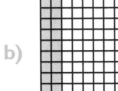

a) $\dfrac{}{100}$ b) $\dfrac{}{100}$ c) $\dfrac{}{}$

2

Write the missing fractions and decimals on these number lines.

a)

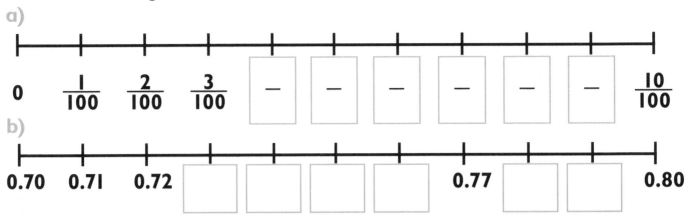

0　　$\dfrac{1}{100}$　$\dfrac{2}{100}$　$\dfrac{3}{100}$　$\dfrac{}{}$　$\dfrac{}{}$　$\dfrac{}{}$　$\dfrac{}{}$　$\dfrac{}{}$　$\dfrac{}{}$　$\dfrac{10}{100}$

b)

0.70　0.71　0.72　☐　☐　☐　☐　0.77　☐　☐　0.80

3

Complete the place value charts for each number.

Example: 0.78

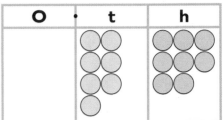

O	.	t	h

a) 0.56

O	.	t	h

b) 0.13

O	.	t	h

c) 0.04

O	.	t	h

Maya needs to gather some sheep. She uses leads, which she has crafted from slimeballs and string, to bring them to her farm. She leads each sheep through the gates and watches as they get used to their new home.

4

Partition these decimals into tenths and hundredths.

Example: 0.28

0.28

0.2   0.08

a) 0.79

b) 0.37

c) 0.62

5

Write the missing numbers in these sequences.

a) 0.34, 0.44, ⬚ , ⬚ , 0.74

b) 0.87, ⬚ , ⬚ , 0.84, 0.83

6

  Answer these by writing decimal numbers in the boxes.

a) Start at 42, count up in tenths.

42, ⬚ , ⬚ , ⬚ , ⬚ , ...

b) Start at 9, count back in hundredths.

9, ⬚ , ⬚ , ⬚ , ⬚ , ...

DIVIDING BY 10 AND 100

With the buildings completed, Maya realises that she needs paths to connect them. While she lays the paths, answer some more fractions and decimals questions.

1

Fill in the spaces to make the sentences true.

a) When 1 is divided by ⬚ , the answer is $\frac{1}{10}$, or 0.⬚ .

b) When 1 is divided by ⬚ , the answer is $\frac{1}{100}$, or 0.⬚ .

2

Show how you can use a place value chart to complete these calculations.

Examples: 84 ÷ 10 = 8.4   32 ÷ 100 = 0.32

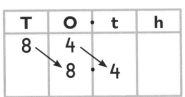   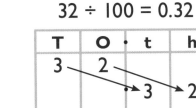

a) 93 ÷ 10 = ⬚

T	O	·	t	h

b) 19 ÷ 100 = ⬚

T	O	·	t	h

c) 7 ÷ 100 = ⬚

T	O	·	t	h

3

Fill in the missing numbers.

a) ⬚ ÷ 10 = 0.6

⬚ ÷ 10 = 0.06

b) ⬚ ÷ 10 = 0.43

⬚ ÷ 10 = 4.3

c) ⬚ ÷ 100 = 0.12

⬚ ÷ 100 = 1.2

d) ⬚ ÷ 100 = 0.04

⬚ ÷ 100 = 0.4

To add some colour and life, Maya plants flowers along the paths and fences. She adds torches on tree logs too, to stop mobs spawning close to her house. She doesn't want creepers blowing up her hard work. Keep up your good work on these questions.

4

Fill in the missing numbers.

a) $83 \div \boxed{} = 0.83$

b) $74 \div \boxed{} = 7.4$

c) $6 \div \boxed{} = 0.06$

d) $47 \div \boxed{} = 0.47$

5

Shade in the two boxes that show a calculation that has an answer with four-tenths.

| $45 \div 10$ | $74 \div 10$ | $43 \div 10$ | $45 \div 100$ | $74 \div 100$ |

6

 Complete each row of the table. The first row is done for you.

Calculation	Picture	Fraction	Decimal
$79 \div 100$		$\dfrac{79}{100}$	0.79
$\boxed{} \div 100$			
$\boxed{} \div 100$			0.68
$\boxed{} \div 100$		$\dfrac{4}{100}$	

COLOUR IN HOW MANY EMERALDS YOU EARNED

COMPARING DECIMALS

The sun is starting to set. Just as Maya prepares to go inside, she notices that a few zombies have appeared in the fields where the sheep are grazing. Grabbing her bow, she takes aim at them from a safe distance.

1

Draw a straight line from Maya's bow to the zombie with the largest decimal number.

Draw a circle around the zombie with the smallest decimal number.

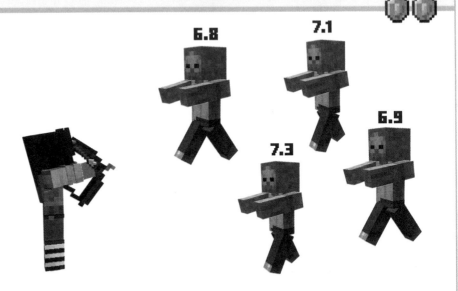

6.8   7.1   6.9   7.3

2

Show how you can use counters in a place value chart to represent each of these decimal numbers.

Example: 6.37

O	.	t	h
⦿⦿⦿ ⦿⦿⦿		⦿⦿⦿	⦿⦿⦿ ⦿⦿⦿ ⦿

a) 4.56

O	.	t	h

b) 9.98

O	.	t	h

c) 9.1

O	.	t	h

d) 8.65

O	.	t	h

e) Use the place value charts to help you order the numbers in parts a), b), c) and d) from smallest to largest.

[ ] < [ ] < [ ] < [ ]

While Maya is fighting the zombies, more mobs begin to spawn. Tomorrow she will put torches in the field – it is too dangerous to do it now. The mobs will not attack her sheep, so she lets them all roam around together and she goes inside.

3

Order these decimal numbers:

a) from smallest to greatest: 9.04, 4.03, 9.84, 5.24, 4.23

b) from greatest to smallest: 5.25, 6.67, 6.53, 4.52, 5.67

4

Here is a table showing the heights of some of Maya's flowers. She says that flower 3 is the tallest and flower 1 is the shortest.

Flower 1	24.34 cm
Flower 2	24.28 cm
Flower 3	23.84 cm

Do you agree? Explain your answer.

5

 Here are some digits on four pieces of carpet in Maya's house: 1   3   5   7

a) Use each of the digits to make this statement true.

5. ☐ ☐ > 5. ☐ ☐

b) Find two more possible answers.

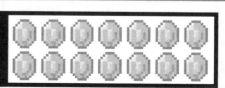

COLOUR IN HOW MANY EMERALDS YOU EARNED

FRACTION AND DECIMAL EQUIVALENTS

Maya's house has two floors. There are four rooms on each floor. However, she forgot to create a room for storage chests. She therefore decides to dig a basement under the house. She will need to remove 100 blocks to create it.

1

Maya's progress in removing the 100 blocks is shown by the fractions below. Draw lines to join each fraction to its decimal equivalent.

| $\frac{1}{100}$ | $\frac{1}{10}$ | $\frac{1}{4}$ | $\frac{1}{2}$ | $\frac{3}{4}$ |

| 0.5 | 0.1 | 0.01 | 0.75 | 0.25 |

2

Write these fractions as decimals.

a) $\frac{8}{10}$ = ☐   b) Six tenths = ☐   c) $\frac{8}{100}$ = ☐

d) $\frac{80}{100}$ = ☐   e) $\frac{32}{100}$ = ☐   f) Ninety-four hundredths = ☐

3

Here is a number line. Write the numbers marked by the arrows as fractions and as decimals.   a)   b)   c)   d)

0

a) Fraction: ☐ — Decimal: ☐   b) Fraction: ☐ — Decimal: ☐

c) Fraction: ☐ — Decimal: ☐   d) Fraction: ☐ — Decimal: ☐

With the basement ready, Maya needs to craft the chests. On the outside of each chest, she places an item frame. She will add an item into the frame to tell her what's inside. During the evening, she sorts all of her materials, tools and other possessions.

4

Maya stops for a break. She drinks 0.3 of a honey bottle.

She says, "0.3 is $\frac{3}{100}$ as a fraction."

Is Maya correct? Explain your answer.

....

5

Here is a number line. Write the numbers marked by the arrows as fractions and as decimals.

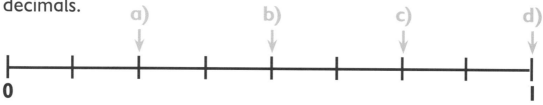

a)   Fraction: ☐ — Decimal: ☐   b)   Fraction: ☐ — Decimal: ☐

c)   Fraction: ☐ — Decimal: ☐   d)   Fraction: ☐ — Decimal: ☐

6

 Maya is filling lots of chests, but some aren't quite full yet.

Write these fractions as decimals.

a) $\frac{2}{4}$ = ☐   b) $\frac{3}{6}$ = ☐

c) $\frac{3}{12}$ = ☐   d) $\frac{150}{200}$ = ☐

ROUNDING DECIMALS

The basement is now nicely organised with rows of storage chests. Some are heavier than others.

1

Here are four decimal numbers: 8.6   8.9   8.1   8.4

a)   Write each number in the correct place on the number line.

8   9

b)   Round each number to the nearest whole number.

8.1 → ☐   8.4 → ☐   8.6 → ☐   8.9 → ☐

2

Here are the masses of some of Maya's storage chests:

14.5 KG   **16.6 KG**   **18.3 KG**   **12.8 KG**   **23.2 KG**   **27.1 KG**

Draw an upwards arrow ↑ on the chests with a mass that rounds **up** to the nearest whole number.

Draw a downwards arrow ↓ on the chests with a mass that rounds **down** to the nearest whole number.

3

Round each number to the nearest whole number.

92.3 → ☐   45.8 → ☐   71.5 → ☐   37.1 → ☐

Before bedtime, there is one more job to do – something Maya has been looking forward to all day! She opens a chest and takes out some items. With everything in her inventory, she wanders over to her old house and begins placing TNT.

4

Maya isn't sure how many blocks make up her old house. It could be 200, 300, or more!

Round each number to the nearest whole number.

345.7 → ☐   388.6 → ☐   268.3 → ☐   210.4 → ☐

5

Maya lights the TNT with flint and steel, then runs to safety about 40 blocks away.

Shade in the boxes with numbers that round to 40 to the nearest whole number.

| 42.8 | 40.9 | 39.8 | 36.8 | 40.2 |

Boom! The TNT explodes. With one bang, it sets off another TNT and the blocks of the old house are all blown to bits.

6

To the nearest whole number, the explosions lasted for 60 seconds.

a) Write down three numbers less than 60 that would round to this number of seconds.

....

b) Write down three numbers greater than 60 that would round to this number of seconds.

....

COLOUR IN HOW MANY EMERALDS YOU EARNED

49

FRACTION AND DECIMAL PROBLEMS

It's dangerous to use **TNT** when mining as it can open holes for water or lava to pour into where you are digging. Destroying a building was the ideal way for Maya to use it. Now it's time for bed.

1

Yesterday, Maya walked 3.7 km. Today, she walked 5.1 km.

Maya estimates how far she has walked in total over the last two days by rounding each distance to the nearest whole number.

Write the total of the rounded distances. ⬜ km

2

The first 8 blocks of coloured wool from Maya's sheep are:

Fill in the boxes in these sentences.

a) $\dfrac{\boxed{}}{8} = \dfrac{\boxed{}}{4} = 0.\boxed{}$

of the wool is blue.

b) $\dfrac{2}{\boxed{}} = \dfrac{1}{\boxed{}} = 0.\boxed{}$

of the wool is red.

3

 Maya had 36 kg of TNT. She used $\dfrac{7}{12}$ of it.

a) How much TNT is left? ⬜ kg

b) The next day, she uses another $\dfrac{3}{12}$ of the TNT. What fraction of the TNT has she used in total now?

$\dfrac{\boxed{}}{\boxed{}}$

c) How much TNT is left? ⬜ kg

COLOUR IN HOW MANY EMERALDS YOU EARNED

ADVENTURE ROUND-UP

OUT WITH THE OLD...

What a tiring but productive day for Maya! After previous adventures, her old house was getting full of things she had collected. There were items and chests everywhere. What she really needed was a new, larger house with more storage space.

...IN WITH THE NEW

Maya first laid the foundations for a new and bigger house. Then, she carefully built up the floors and constructed the rooms. She transformed the outside land — first creating a garden to grow crops, then building a barn and a grazing area for sheep.

HOME SWEET HOME

Once Maya had finished the building work, she sorted out all of her materials, tools and other possessions. She ended up with chests full of stone, some full of wood and others filled with ore and gems. After digging out a basement to store the chests, she added item frames to them as signs. Everything now has its own place and Maya has the space she needs.

MEASUREMENT

JUNGLE HILLS

The jungles of the Overworld are difficult to navigate simply because of the tightly-packed trees. In jungle hills, exploring is made harder by steep slopes and mountains. However, the jungle tree is the only plant to grow cocoa beans. You can also find yummy melon sprouting in patches.

FULL OF LIFE

The jungle is bursting with colour and life. Parrots come in several colours and like to eat seeds, then travel with adventurers. Pandas love rolling around and chomping bamboo. Ocelots are shy but can be seen scampering through the grass.

DON'T FALL INTO A TRAP

Mobs love to hide under the leaves or behind a tree trunk to catch a hero by surprise. Tucked away in some jungles, jungle temples can be found. Hidden treasure can be the reward for anyone who doesn't trigger any traps!

HIGH AMBITIONS

Oscar is standing on top of a large hill in the jungle with views in all directions. Even up here, the trees grow tall. He is here to study the plants, watch the animals and collect resources. Oscar has brought his pet llama to help carry the goods back home.

ESTIMATING MEASURES

These jungle hills border an ocean on one side. Oscar has also seen several small ponds among the trees.

 1

Draw lines to match the best estimate for the volume of each item.

Bathtub	Sink	Jug	Small pond

300 l	15 l	15,000 l	1 l

2

a) Circle the best estimate for the height of a llama.

2 m   **2 mm**   **200 m**   **20 km**

b) Circle the best estimate for the length of a cod fish.

50 cm   **50 m**   **5 km**   **500 cm**

3

The jungle is one of the warmer biomes.

Estimate the temperature shown on the thermometer.

0   20   40   60 °C

☐ °C

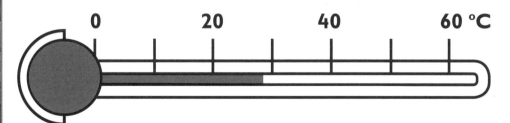

CONVERTING AND COMPARING MEASURES

| 100 cm = 1 m | 1,000 m = 1 km | 1,000 ml = 1 l | 1,000 g = 1 kg |

Oscar travelled far from his home in the badlands to reach the jungle. He lost count of how many kilometres he trekked and how many litres of water he drank. Thankfully he had his llama to carry the many kilograms of equipment and supplies he needs. Oscar ties his llama to a wooden post before heading deeper into the jungle.

1

Convert these measurements.

a) 900 cm = [   ] m

b) 2 m = [   ] cm

c) 8,000 m = [   ] km

d) 7 km = [   ] m

e) 3,000 ml = [   ] l

f) 7 l = [   ] ml

g) 3 kg = [   ] g

h) 4,000 g = [   ] kg

2

Convert these times.

a) 360 seconds = [   ] minutes

b) 10 minutes = [   ] seconds

c) 3 years = [   ] months

d) 28 days = [   ] weeks

As Oscar steps through the tree trunks, he emerges into a small clearing. Just ahead are a parrot, an ocelot and a panda. Oscar chops some bamboo and feeds the panda. He harvests seeds from tufts of grass to feed the parrot, but the ocelot runs away when he approaches it with raw fish.

 3

Solve these problems.

a) Three fallen branches weigh 400 g, 825 g and 375 g.

What is the total mass of all three branches in kilograms? ☐ kg

b) Three bamboo sprouts have lengths of 250 cm, 125 cm and 425 cm.

What is their total length in metres? ☐ m

c) Four trailing vines have lengths of 5 m, 450 cm, 320 cm and 8 m.

Sort these measurements from shortest to longest.

....

 4

Write <, > or = in the boxes to make each statement true.

a) 7 kg ☐ 4,000 g   b) 2,000 m ☐ 5 km

c) 19 m ☐ 18 km   d) 4,000 ml ☐ 4 l

 5

 A bucket holds 5 litres of water. It is used to fill one container that holds 1,500 ml and a second container that holds twice as much as the first.

How much water is left in the bucket after both containers are filled?

☐ ml

PERIMETER

The small clearing seems like such a peaceful spot. The trees aren't so close together here, so the light from the sun warms the grass. Oscar could sit here for a long time in the quiet, but he wants to find out where that ocelot went.

1

Find the perimeter of this section of the jungle floor.

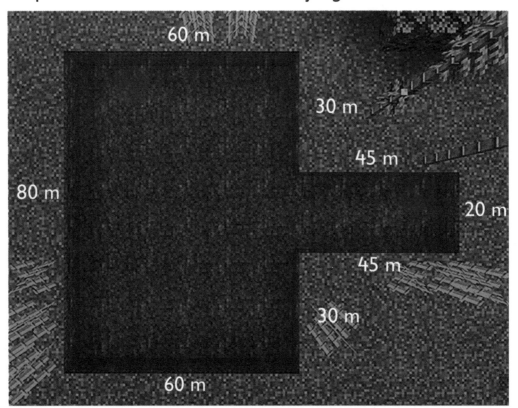

60 m

30 m

45 m

80 m

20 m

45 m

30 m

60 m

□ m

2

Measure each side of this shape with a ruler.

Find the perimeter of the shape.

□ cm

Oscar finds the ocelot hiding behind a tree. It is very shy and wary of him. He crouches down with the raw fish in his hand and creeps very slowly. The ocelot looks out at him and steps forwards, sniffing at the fish. It soon comes closer and takes the fish happily.

3

The perimeter of this rectangle is **38 cm**.

What is the length of the rectangle?

[ ] cm

7 cm

4

The perimeter of this pond is **370 m**.

Calculate the missing length.

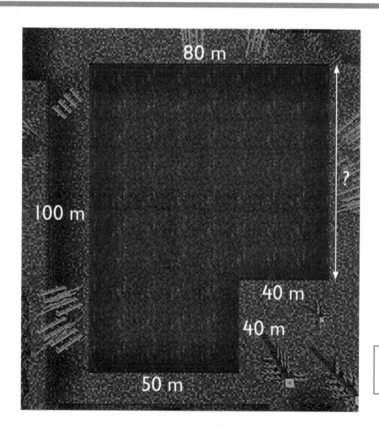

80 m

100 m

?

40 m

40 m

50 m

[ ] m

5

 High above the jungle, a cloud drifts through the sky. It looks like it is made from rectangles that are 3 km long and 1 km wide.

1 km

3 km

What is the perimeter of the cloud?

[ ] km

AREA

Continuing through the jungle, Oscar discovers a crop of melons. When he harvests the melons, they break into slices. He eats a few to restore his health and keeps others for the seeds, which he will plant at home.

1

Grids A, B, C and D show four plots that are being used to grow melons. Each small square on the grids is 1 m². Each green square is being used to grow melons.

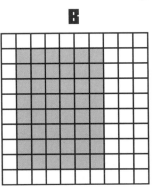
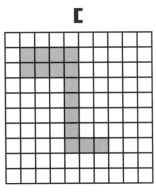
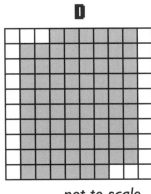

not to scale

a) Find the area being used to grow melons in each plot.

A = ☐ m²   B = ☐ m²   C = ☐ m²   D = ☐ m²

b) Sort plots A to D from the smallest to the largest area of melon crop.   ....

2

Oscar would like to grow melons at home. To design the shape of his plot, use this grid that has squares of area 1 cm².

a) Oscar wants the plot to cover 10 squares of the grid.

Using the grid, draw two different shapes of plot that he could create.

b) What is the area of each of the shapes you have drawn? ☐ cm²

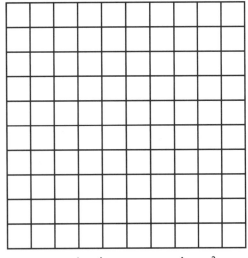

scale: 1 square = 1 cm²

Oscar plans to split his farm at home into sections, each inspired by different biomes. For this he will need lots of materials from the jungle hills, so he collects jungle logs, vines and clumps of leaves.

3

Here is a plan of a jungle garden at the entrance to Oscar's farm. Each square is 1 m².

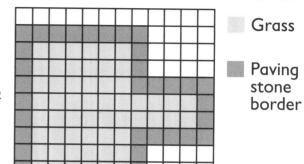

Grass

 Paving stone border

a) What is the area of the grass section? ___ m²

b) In pounds, each 1 m² of grass costs £4.

How much will the grass cost for this garden? £ ___

c) What is the total area of the paving stone border? ___ m²

d) Each paving stone is 1 m². In pounds, they each cost £15.

How much will the paving stone border cost? £ ___

e) What is the total cost of the grass and the paving stones? £ ___

4

Here are two designs, A and B, for paddocks on Oscar's farm.

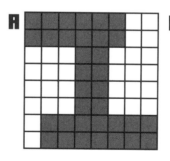

a) Draw a shape that has an area greater than A but less than B.

b) Draw a shape that has the same area as A but is symmetrical.

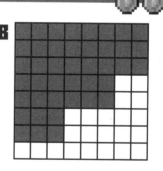

COLOUR IN HOW MANY EMERALDS YOU EARNED

MONEY

Oscar's inventory is getting full. Checking his compass, he heads back in the direction of his llama.

1

Write the amounts of money shown.

Example: £5 and 32p

£5.32

a)

£ ▢ and ▢ p

£ ▢

b)

£ ▢ and ▢ p

£ ▢

c)

£ ▢ and ▢ p

£ ▢

d)

£ ▢ and ▢ p

£ ▢

2

Order these amounts of money from smallest to greatest:

   £5.32   £10.87   527p   1184p   £7.89

....

Oscar is eager to return home and begin crafting and building. He has a lot of slices to make glistering melon and a stack of jungle logs to start a cocoa bean farm. Once the cocoa beans are ready to harvest, he will make lots of cookies and sell them for a handsome price!

3

Write **<**, **>** or **=** in the boxes to make each statement true.

a) £6.89 ☐ 680p

b) 584p ☐ £3.25

c) 872p ☐ £12

d) £9.87 ☐ 987p

4

 Here are the prices in pounds and pence of some items at a food market:

Melon:
£3.25

Cookie:
£2.30

Glistering melon:
£4.80

Cocoa beans:
£1.65

a) A customer buys three melons and pays with a £10 note.

How much change do they get? ☐

b) Another customer buys one glistering melon and one cookie. They pay with a £10 note.

How much change do they get? £ ☐

c) A third customer wants to buy three glistering melons and one lot of cocoa beans. They have £15.

How much more money do they need? £ ☐

12- AND 24-HOUR TIME

Checking the time, Oscar sees that it is getting late. The sun is starting to set behind the trees on the mountain. He can already see a few mobs wandering between the trees. Every few steps, Oscar fires an arrow into a creeper or tackles a zombie with his sword.

Write down the times shown on these clocks in words.

a)

....

b)

....

c)

....

2

Convert the 12-hour times to 24-hour times.

a) 4:35 am ....   b) 4:28 pm ....

c) 12:37 am ....   d) 12:17 pm ....

e) 3:52 am ....   f) 7:26 pm ....

Oscar doesn't want to spend much more time in the jungle. The mobs are getting very mischievous and, in the failing light, it's hard to tell what's a tree and what's a skeleton!

3

Convert the 24-hour times to 12-hour times.

a) 20:08 ....

b) 03:54 ....

c) 23:18 ....

d) 07:52 ....

e) 09:56 ....

f) 17:56 ....

4

It is the afternoon. A clock shows this time:

a) Write the time in words.

....

b) Write the time in 12-hour time. ....

c) Write the time in 24-hour time. ....

5

It is the morning. A clock shows this time:

a) Write the time in words.

....

b) Write the time in 12-hour time. ....

c) Write the time in 24-hour time. ....

COLOUR IN HOW MANY EMERALDS YOU EARNED

TIME PROBLEMS

Oscar runs as fast as possible, dodging everything in his way. Arrows whizz past his head and creepy arms reach out from behind trees. He hears an Enderman following him. Finally, he reaches his patient llama before grabbing the lead and dashing off into the dusk.

1

Oscar arrives home at 22:15. It takes him 90 minutes to unpack the chests.

a) Draw the time on the analogue clock to show when Oscar finishes unpacking.

b) Write the time that he finishes unpacking in 12-hour time. ....

c) Write the time that he finishes unpacking in 24-hour time. ....

2

Write **<**, **>** or **=** in the boxes to make each statement true.

a)   4 days   ☐   48 hours

b)   3 weeks   ☐   25 days

c)   2 years   ☐   700 days

d)   50 months   ☐   5 years

3

 Oscar notices that four villagers visit a neighbour's house that evening.

Villager A arrives at	Villager B arrives at	Villager C arrives at	Villager D arrives at

and stays 57 minutes.	and stays 43 minutes.	and stays 35 minutes.	and stays 22 minutes.

Which villager leaves the house first? ....

COLOUR IN HOW MANY EMERALDS YOU EARNED

ADVENTURE ROUND-UP

HILLS TO HOME WITH A HAUL

Oscar's llama is back in the warmth of its barn. It wasn't until he unloaded the chests that he realised how many things he had collected in jungle hills. There are logs, planks, food, vines, leaves... the list goes on! If only he could have brought home a big, cuddly panda!

GOODNIGHT, OSCAR

After such a long journey, Oscar needs some rest. He refreshes with a drink of milk and an apple, but it's now very late and time for bed. As he lays down his head on the bed, he begins to dream of turning his planned farm into reality.

GEOMETRY

THE WASTES AWAIT

Unlike other biomes in the Nether, the Nether wastes have few features. There are no trees, no basalt columns and very few piglins, though you may find a fortress. Zombie pigmen wander the wastes looking for something… but nobody knows what.

MATERIAL RICHES

The deep red of the netherrack is everywhere – on the ground, on walls and even on ceilings. Adventurers looking to get rich may be able to harvest Nether gold. Builders come here for Nether quartz and glowstone.

MIND YOUR STEP

The wide, flat spaces of the Nether wastes mean that a brave explorer can create a base here. That's as long as they don't plan to sleep… beds will explode when used! Everyone must mind their step; there are slopes that often end in a lake of lava.

THE WAY TO THE WITHER

Maya has come to the Nether wastes for one reason. She has collected three Wither skulls and plans to conjure the Wither – a floating three-headed boss mob – for a big showdown. Maya has brought some building materials to create an arena, where the fight will be held.

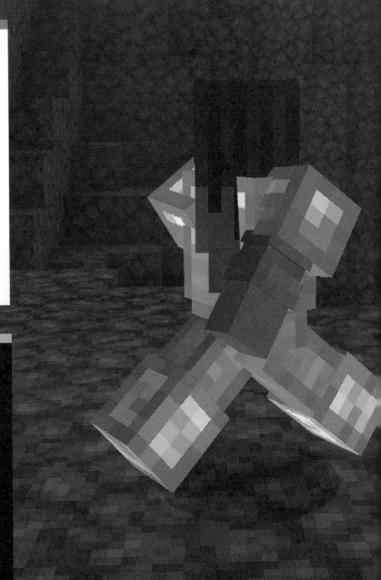

ANGLES

Maya's arena will need plenty of places to hide and some raised areas from which arrows can be fired. She starts by creating the shape of the arena. She experiments with different angles, to create hidey-holes.

1

Here are some angles that Maya has used:

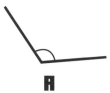  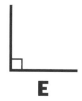

A   **B**   **C**   **D**   **E**

a) Label each angle as **acute**, **obtuse** or a **right angle**.

A ....   B ....   C ....

D ....   E ....

b) Order the angles in size from smallest to largest. ....

c) Draw an angle that is smaller than the smallest angle.

d) Draw an angle that is larger than the largest angle.

2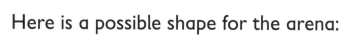

Here is a possible shape for the arena:

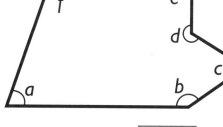

a) Circle the acute angles in the shape.

b) Write **>**, **<** or **=** in the boxes to make each statement true.

angle *a* ☐ angle *e*   angle *e* ☐ angle *f*   angle *b* ☐ angle *f*

TRIANGLES

Triangular shapes are difficult to build, but Maya would like to use them in her arena as they might confuse the Wither. With the right protection overhead, they make great areas to hide while drinking a potion or regenerating health with a quick snack.

1

Here are some triangles:

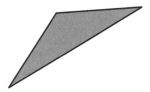

....

....

a) Label each triangle as **equilateral**, **isosceles** or **scalene**.

b) Circle any triangles that have an obtuse angle.

c) Tick any triangles that have a right angle.

2

a) Draw a right-angled triangle on the grid.

b) Draw an isosceles triangle on the grid.

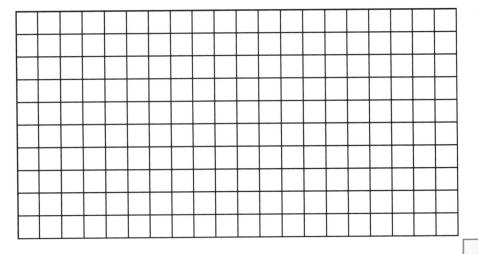

Maya begins building up the sides of the arena. She plans to spawn the Wither in the very centre of the arena, then dash up to a higher platform and begin firing arrows from different angles.

3

Put a tick in each row to show whether each statement is **always true**, **sometimes true**, or **never true**.

Statement	Always	Sometimes	Never
A triangle has one obtuse angle.			
A triangle has at least one acute angle.			
A triangle has three acute angles.			
A scalene triangle has two sides the same length.			
All three sides of a triangle are the same length.			

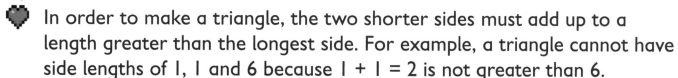

4

 In order to make a triangle, the two shorter sides must add up to a length greater than the longest side. For example, a triangle cannot have side lengths of 1, 1 and 6 because 1 + 1 = 2 is not greater than 6.

a) Decide whether you could make a triangle using each set of side lengths shown in the table. Write **Yes** or **No** in the final column. The first one has been done for you.

Side 1	Side 2	Side 3	Triangle?
1	1	6	No
1	2	5	
1	3	4	
2	2	4	
2	3	3	

b) If you answered 'Yes' in any row above, is the triangle isosceles, scalene or equilateral?

....

QUADRILATERALS

It is a little harder to build in the Nether because there are mobs wandering about and creating mischief – especially the ghasts, which blow holes into Maya's structure with their fireballs. But at least that's good practice for the Wither.

1

Tick the shapes that are quadrilaterals.

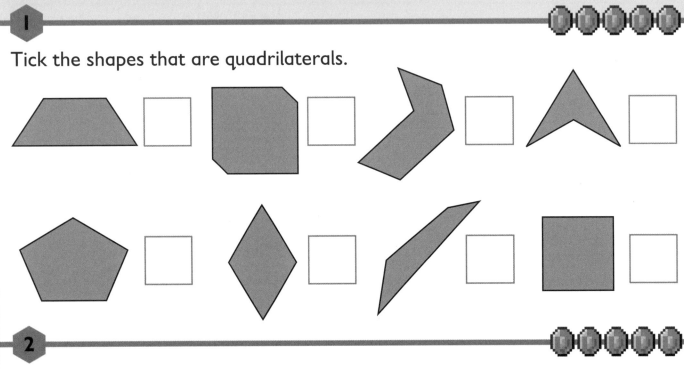

2

Draw lines to join each quadrilateral to its correct name.

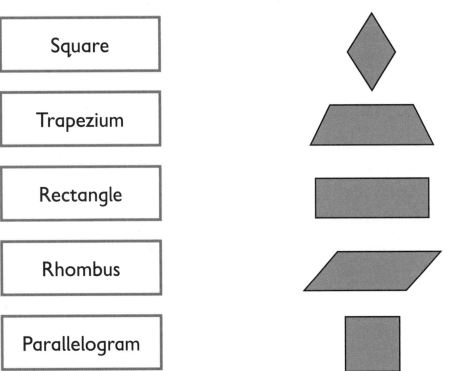

Square

Trapezium

Rectangle

Rhombus

Parallelogram

The arena is almost finished. Maya hasn't put too much time into the construction because she knows a lot of it will be destroyed.

3

Draw examples of each of these shapes on the grid.

Parallelogram   **Rectangle**   **Rhombus**   **Square**   **Trapezium**

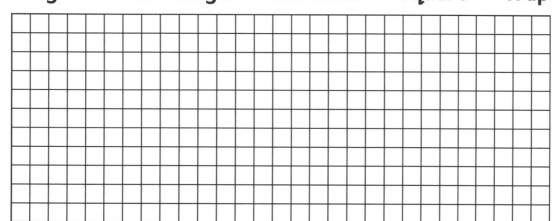

4

Fill in the table to show where each of these quadrilaterals fit:

Parallelogram   **Rectangle**   **Rhombus**   **Square**   **Trapezium**

	Two pairs of parallel sides	Exactly one pair of parallel sides
Always four equal sides		
Not always four equal sides		

COLOUR IN HOW MANY EMERALDS YOU EARNED

LINES OF SYMMETRY

It's time to summon the **Wither**. **Maya** leaves the arena and places a chest next to the **Nether** portal. In the chest she places the last of her building materials and any items not needed for the fight. **Maya** walks back to the centre of the arena. She places soul sand in the shape of a 'T' and puts two of the **Wither** skulls on top. After placing the final skull, she runs.

1

Draw the lines of symmetry on these shapes.

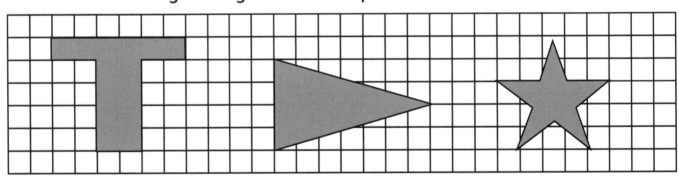

2

The red dashed lines are lines of symmetry. Complete the shapes.

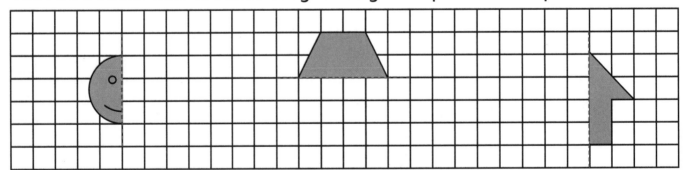

3

How many lines of symmetry do each of these shapes have?

a)   b)   c)

....

The Wither bursts into life behind Maya. An explosion rips through the arena and blasts her backwards! The Wither spots Maya running away and begins launching blazing skulls towards her.

4

The red dashed lines are lines of symmetry. Complete the symmetrical designs of the Wither and a Wither skull.

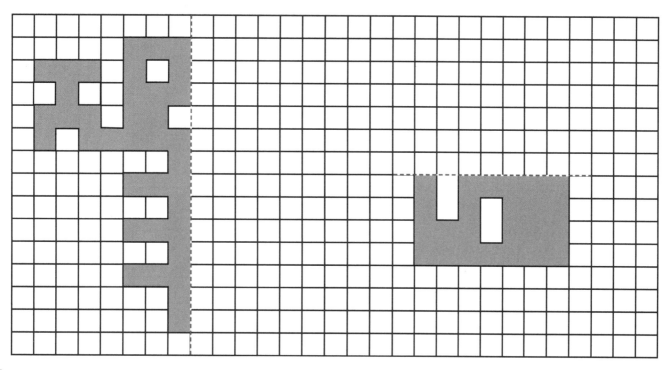

5

 Draw three more lines to create a pentagon with one line of symmetry.

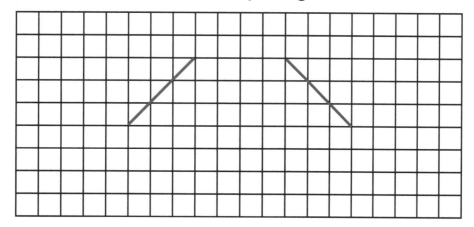

CO-ORDINATES

The **Wither** flies menacingly around the arena. **Maya** fires arrows at it from a safe distance and, when the **Wither** seems confused or stops for a moment, she dashes in and whacks it with her netherite sword.

1

These points show the position of Maya as she dashes around the Wither.

Write the co-ordinates of each point.

A (............ , ...)   B (............ , ...)

C (............ , ...)   D (............ , ...)

E (............ , ...)   F (............ , ...)

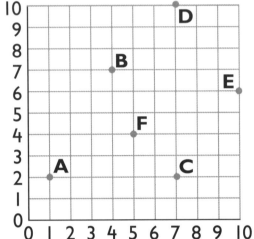

2

The skulls fired by the Wither land on the co-ordinates shown on the grid.

Write down the letter that matches the co-ordinates to find the answer to these two jokes.

a) What do you get if you cross a sheep with a zombie?

(1, 1)   (4, 7) (1, 1) (1, 1) (1, 1) (7, 0)

☐   ☐ ☐ ☐ ☐ ☐

(4, 1) (7, 10) (6, 3)

☐ ☐ ☐

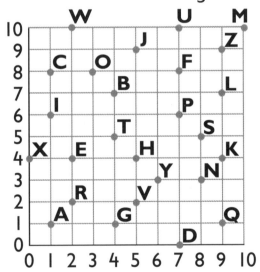

b) Why did the llama cross the road?

(1, 6) (4, 5)   (2, 10) (1, 1) (8, 5)   (4, 5) (5, 4) (2, 4)

☐ ☐   ☐ ☐ ☐   ☐ ☐ ☐

(1, 8)(5, 4)(1, 6)(1, 8)(9, 4)(2, 4)(8, 3)(8, 5)   (7, 0)(1, 1)(6, 3)   (3, 8)(7, 8)(7, 8)

☐ ☐ ☐ ☐ ☐ ☐ ☐ ☐   ☐ ☐ ☐   ☐ ☐ ☐

One of the Wither skulls hits Maya in the back. Suddenly, she is overcome by the Wither effect. It feels like being poisoned and her health starts to fall. She needs to find a safe place to heal up. A blue skull then flies towards her. Maya swings her sword and hits it back towards the Wither!

3

After quickly drinking milk and healing, Maya finds more blue skulls to fire back at the Wither. The skulls are found at the co-ordinates written below. Mark each point on the grid and write the capital letter beside it.

A (3, 9) B (4, 0) C (6, 2)

D (10, 6) E (1, 5) F (8, 3)

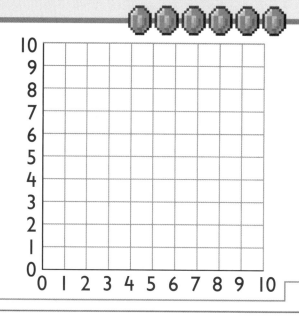

4

Use the clues to write down and plot the points for A, B, C and D.

a) My x co-ordinate is twice my y co-ordinate.

My y co-ordinate is 3 less than 7.

A (.........., ...)

b) My x co-ordinate is an even number between 5 and 9.

My x co-ordinate is less than 8.

My x co-ordinate is three times my y co-ordinate. B (.........., ...)

c) My x co-ordinate and y co-ordinate add up to 6.

My x co-ordinate is 2 greater than my y co-ordinate. C (.........., ...)

d) My x co-ordinate and y co-ordinate add up to 15. Both co-ordinates are a multiple of 3. My y co-ordinate has the greater value.

The difference between my x and y co-ordinates is 3. D (.........., ...)

COLOUR IN HOW MANY EMERALDS YOU EARNED

TRANSLATION AND SHAPES ON A GRID

Maya feels refreshed and confident again. The blue skulls fly through the air and strike the Wither. She wants to get in close and use her sword. Thankfully, she placed random blocks of netherrack on the floor of the arena. She uses these as cover, trying to hide and catch the Wither by surprise.

1

Shapes A to F show blocks of netherrack on a grid.

Write the letter that each block moves to after these translations.

Example: *Block A moves 3 right and 1 down.*

a) Block A moves 1 left and 3 down

b) Block C moves 1 right and 2 down

c) Block E moves 4 left and 1 down

d) Block F moves 3 right and 4 up

| **B** |
| |
| |
| |
| |

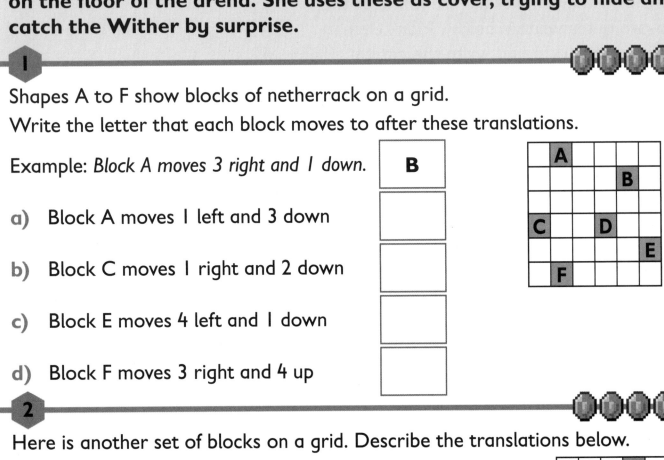

2

Here is another set of blocks on a grid. Describe the translations below.

a) Block A to Block B ....

b) Block F to Block D ....

c) Block C to Block E ....

d) Block E to Block B ....

3

A trapezium is drawn on a grid. It moves 3 left and 2 up.

Write down the translation that will move it back to its original position.

....

The Wither creates a large explosion. Suddenly, three Wither skeletons appear, and they are after Maya! She splashes herself with a Potion of Strength. Moving quickly, she destroys the Wither skeletons before striking the Wither over and over again with her sword. The Wither is weakened. Maya dodges the fast-moving black skulls and bats back the slower blue skulls.

4

Maya's positions are represented by the points on this co-ordinate grid.

Below are some translations of the points. Write the letter of the position where the point moves to.

Example: *Point A moves 2 right and 2 up.* **F**

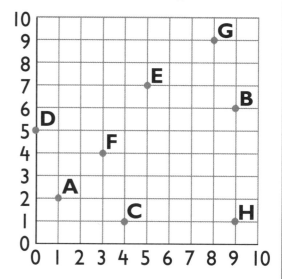

a) Point C moves 1 left and 3 up

b) Point E moves 4 right and 1 down

c) Point B moves 1 left and 3 up

d) Point H moves 6 left and 3 up

5

The Wither is confused and in danger. Its movements become easy to follow.

a) The co-ordinates below represent positions that the Wither is seen in.

Mark the points on the co-ordinate grid. Connect the points with straight lines to create a shape (join point A to B, point B to C, point C to D and point D to A).

A (1, 4) B (2, 6) C (6, 6) D (8, 4)

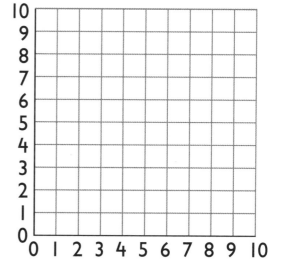

b) What shape have you drawn? ....

Maya hits back a blue skull and darts behind the **Wither**. She holds her sword up high and brings it crashing down. The **Wither** spins on the spot. The sword strikes again and the **Wither** is defeated in an explosion of smoke and particles!

6

A co-ordinate grid is shown.

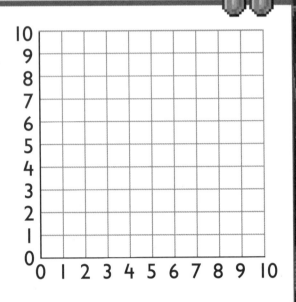

a) Plot the points A (1, 4), B (3, 6) and C (8, 6) on the co-ordinate grid. These are three vertices of a parallelogram.

b) Plot the fourth vertex of the parallelogram and label it as point D.

What are the co-ordinates of point D?

(............, ...)

7

There are several experience orbs for Maya to collect.

a) Draw these points on the co-ordinate grid as colourful experience orbs. Connect the points with straight lines to create a shape (join point A to B, point B to C, point C to D and point D to A).

A (2, 3) B (3, 1) C (4, 3) D (3, 5)

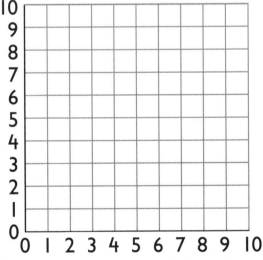

b) What shape have you drawn?

....

c) Translate the shape 3 right and 2 up and draw it on the grid. Label the vertices E, F, G and H (in the same order as points A, B, C and D on the first shape).

d) Write down the co-ordinates of the translated points.

E (............, ...) F (............, ...) G (............, ...) H (............, ...)

ADVENTURE ROUND-UP

A BATTLE ROYAL

That was a battle to end all battles! The arena is a mass of destruction, with blocks littering the area. The arena did its job, holding the Wither in place, confusing it with tight spaces and providing lots of cover to hide behind.

PREPARATION PAYS OFF

There were many dangerous moments for Maya. Getting hit by a skull which caused her health to drain away was particularly scary. Luckily, she had buckets of milk which can cure the Wither effect. Everything in the fight came down to planning — she chose the right place and the best equipment. She used all of her arrows and potions. She's very tired, but she's still standing!

STATISTICS

BEAUTY OF THE BADLANDS

The badlands could be described as barren, yet beautiful. The sand is a deep orange colour which breaks apart as the hills and mountains rise to reveal a stunning view. Below the sand are layers of terracotta, which comes in several colours, from a dark chocolate brown to a light chalky white.

BUILDER'S PARADISE

The badlands biome is the perfect place to harvest building materials. The terracotta breaks easily with a pickaxe and can be used just as it is. From a distance, the layers that build up the tall peaks of the badlands are a precious sight. More precious is what can be found deep underground.

GOLD RUSH

In other biomes, an adventurer is lucky to find a patch of gold when mining. In the badlands, you cannot dig for long without finding the much-desired ore!

WHAT'S IN STORE?

Oscar has set up a small shop near his home in the badlands. He has a mine that keeps producing gold, which he trades in the Nether for rare items. He stocks these items in his shop, along with homemade food.

PICTOGRAMS

Oscar sells all kinds of food to those who pass by. One of the best sellers is the cake. He sells several cakes every day – sometimes the whole cake, sometimes in slices.

 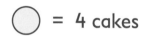

This pictogram shows the number of cakes that Oscar sells each day.

Monday	◯ ◯ ◯ ◯
Tuesday	◯ ◯ ◯ ◯ ◯ ◖
Wednesday	◯ ◯ ◯ ◖
Thursday	◯ ◯
Friday	

◯ = 4 cakes

a) How many cakes are represented by each full circle?

b) How many cakes did Oscar sell on Monday?

c) How many more cakes did he sell on Tuesday than on Monday?

d) On Friday, he sold 12 cakes.
    How many circles will complete the pictogram for Friday?

e) On which day did Oscar sell fewest cakes? ....

f) On how many days did Oscar sell more than 12 cakes?

g) How many cakes in total did he sell from Monday to Friday?

BAR CHARTS

One of the other things that Oscar sells is tools. Some people, way out in the badlands, like to buy pickaxes or swords. Oscar enchants them, then sells them to anyone who is going on a mining expedition. He has several which sell very well.

1

Use this bar chart to answer questions 1–5. It shows the number of tools that Oscar sold in one week.

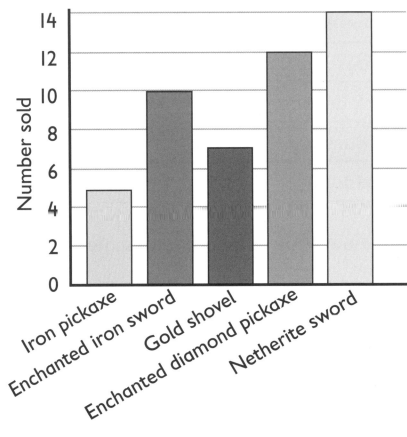

Tools Sold in One Week

a) What is the value of each interval on the vertical axis?

b) How can you work out the exact number of iron pickaxes sold?

....

2

a) Which tool did Oscar sell the most? ....

b) Which tool did he sell the least? ....

c) How many more of the best-selling tool were sold than the least popular tool?

At the end of every week, when the store is closed, Oscar crafts stacks of new tools. He creates them, then enchants them using his experience and stacks of lapis lazuli. During the course of his adventures, he has become an expert on what tool is best for the job.

3

Look again at the bar chart on page 82.

a) Which tool did Oscar sell 7 times? ....

b) How many tools did he sell more than 9 times?

4

Complete the table using the information in the bar chart.

Tool	Number sold
Iron pickaxe	
Enchanted iron sword	
Gold shovel	
Enchanted diamond pickaxe	
Netherite sword	

5

How many tools did Oscar sell in total during the week?

COLOUR IN HOW MANY EMERALDS YOU EARNED

TIME GRAPHS

::

The badlands is a warm biome but it turns cooler at night.

1

Oscar sometimes measures the outside temperature. This time graph shows the temperature from one particular evening to the next morning. Use it to answer questions 1 and 2.

Overnight Temperature

a)   At what time did Oscar first record the temperature? ....

b)   What was the lowest temperature he recorded? ☐ °C

c)   At what time was the lowest temperature recorded? ....

d)   At what time did the temperature reach 10°C? ....

2

Look again at the time graph above.

a)   What is the difference between the highest and the lowest temperatures recorded? ☐ °C

b)   How many hours did it take from the first recording for the temperature to fall to 2°C? ☐ hours

c)   i)   Estimate the temperature at midnight. ☐ °C

   ii)   Explain how you estimated it. ....

....

3

Oscar recorded the height of a bamboo plant for six weeks.

Week	Week 1	Week 2	Week 3	Week 4	Week 5	Week 6
Height (cm)	4	6	10	11	13	16

a) Create a time graph to display the heights he recorded.

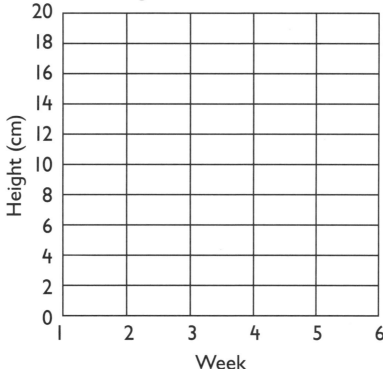

Height of Bamboo Plant

b) Between which two weeks did the plant reach a height of 12 cm?

....

PRESENTING DATA

Oscar has a visitor. It's his friend, Maya! While chatting in the shop, they discuss how many mobs they've seen and fought together on past adventures. Oscar surprises her by saying that he kept a list of them.

 1

This table and pictogram show information about five mobs that Oscar and Maya have defeated together.

Spiders	800
Zombies	1,200
Skeletons	900
Creepers	
Drowned	

Spiders	● ● ● ●
Zombies	
Skeletons	● ● ● ● ◖
Creepers	● ● ● ◖
Drowned	● ● ● ● ●

Key: ● = [ ] mobs

a) How many mobs does each full circle in the pictogram represent? Complete the pictogram's key.

b) Complete the table.

c) Complete the pictogram.

2

Use the table and pictogram in question 1 to answer these questions.

a) Which mob have Oscar and Maya defeated the most number of times? ....

b) Which mob have they defeated the fewest number of times? ....

c) How many mobs have they defeated in total? [ ]

d) What is the difference in the number of creepers and zombies defeated? [ ]

Oscar and Maya talk until the sun starts to set. Oscar closes the door and the pair sit in his kitchen. Maya will stay for dinner before her journey home. Oscar cooks chicken and baked potatoes while they carry on talking about mobs. Maya likes defeating creepers because they drop gunpowder. Oscar likes defeating skeletons because he can collect arrows for his bow.

3

 Follow these steps to create a bar chart to show the information in question 1.

a) Write suitable labels for the horizontal axis and the vertical axis in the correct place on the bar chart. Also write a suitable title for the chart.

b) By how much should the values on the vertical axis increase at each interval? Circle the best answer.   **200**   **50**   **1,000**   **I**   **2**

c) Write the scale on the vertical axis.

d) Label each category on the horizontal axis.

e) Draw the bars on the chart.

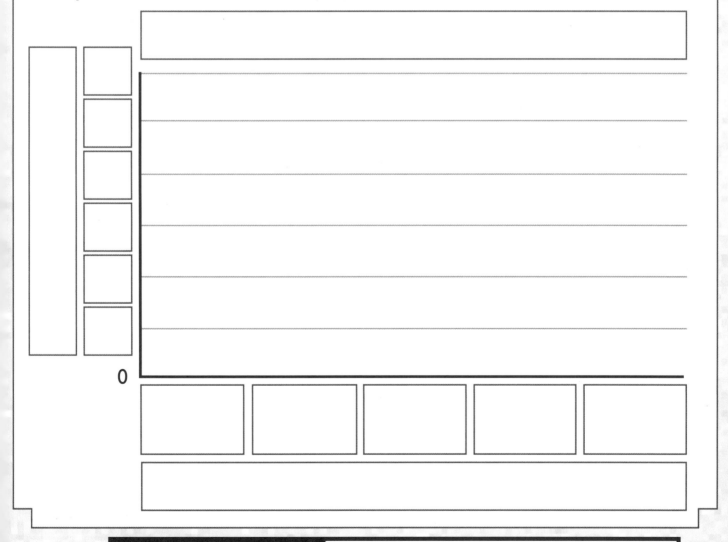

COMPARING DATA

Maya is ready to go home. She thanks Oscar for the lovely dinner. Maya climbs on her horse and rides off towards home. Oscar feeds his animals and enjoys the sunset before bedtime.

1

Here is a bar chart showing the number of different animals on Oscar's farm.

a) Which animal does Oscar have most of?

....

b) How many sheep does he have?

c) How many more chickens than cows does Oscar have?

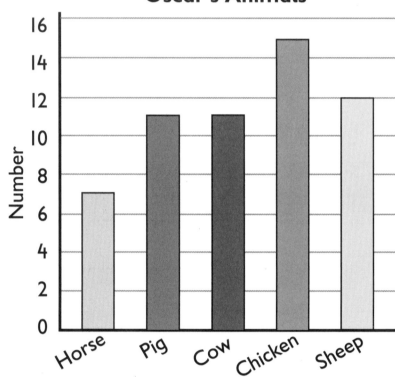

Oscar's Animals

d) How many pigs and horses does he have altogether?

2

Complete these sentences about the data shown in question 1.

a) Oscar has ☐ animals in total.

b) Oscar has the same number of .... and ....

c) $\frac{1}{8}$ of the animals are ....

d) Oscar has 8 more .... than ....

COLOUR IN HOW MANY EMERALDS YOU EARNED

ADVENTURE ROUND-UP

SHOPKEEPING SUCCESS

Life is quite different for Oscar now. The shop has become so popular that he has less and less time for exploring. He enjoys meeting the customers and crafting lots of cool items to sell. As well as that, he still has a farm to run and a mine to keep working.

MORE MEMORIES TO BE MADE?

Oscar enjoys visits from his friend, Maya. They like to share memories of their victories over hostile mobs, the weird and wonderful biomes they have visited, and the loot they have plundered. Perhaps there are more adventures still to come...but that's for another time.

ANSWERS

Page 5

1. a) 350   550   [1 emerald]
   b) 75   125   [1 emerald]
   c) 27   45   [1 emerald]
   d) 12   18   [1 emerald]
2. a) Add 7   b) Add 6
   c) Subtract 25   d) Subtract 9   [1 emerald each]
3. Grid completed from left to right: 38; 37; 57; 56
          [1 emerald each]

Pages 6–7

1. a) 6   b) 4   c) 3   [1 emerald each]
2. a) seven   b) seven thousand
   c) seven hundred   d) seventy   [1 emerald each]
3. Largest no.: 9,762   Smallest no.: 2,679 [1 emerald each]
4. a) >   b) >   c) <   d) <   [1 emerald each]
5. 8,023   8,467   11,569   13,667   17,421 [1 emerald]

Pages 8–9

1. a) 3,250   5,250   [1 emerald]
   b) 4,465   7,465   [1 emerald]
   c) 7,852   5,852   [1 emerald]
   d) 5,648   4,648   [1 emerald]
2. a) 5,237   b) 335
   c) 3,012   d) 4,066   [1 emerald each]
3. a) 5,758   b) 9,301
   c) 4,437   d) 9,576   [1 emerald each]
4. Table completed by each row from the top:
   2,264   4,264   [1 emerald]
   1,621   3,621   [1 emerald]
   8,573   9,573   [1 emerald]
   7,482   8,482   [1 emerald]
5. a) Chest A: 4,537   Chest B: 4,587
   Chest C: 6,594   Chest D: 5,862   [1 emerald each]
   b) 6,594   5,862   4,587   4,537   [1 emerald]

Pages 10–11

1. a) 1,236   [1 emerald]
   b)

   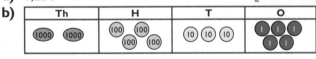

Th	H	T	O

          [1 emerald]

   c)

   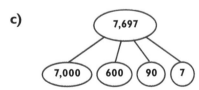

          [1 emerald]
   d) 4,000 + 500 + 70 + 9   [1 emerald]
2. Numbers placed on number line from left to right:
   1,203   2,016   2,615   3,591   [1 emerald]
3. a) Two thousand, three hundred and twenty-two
   b) Four thousand, one hundred and ninety-five
          [1 emerald each]

4. a) 5,577   b) 3,063   c) 4,406   [1 emerald each]
5. Any suitable estimates from the left:
   350–450   1,200–1,300
   1,850–1,950   2,700–2,800   [1 emerald each]

Pages 12–13

1. a) Numbers placed on number line from left to right:
   209   214   242   288   [1 emerald]
   b)

Number	Nearest 10	Nearest 100
242	240	200
288	290	300
214	210	200
209	210	200

        [1 emerald for each row]
2. a) Numbers placed on number line from left to right:
   6,021   6,384   6,782   6,828   [1 emerald]
   b)

Number	Nearest 10	Nearest 100	Nearest 1,000
6,384	6,380	6,400	6,000
6,828	6,830	6,800	7,000
6,782	6,780	6,800	7,000
6,021	6,020	6,000	6,000

        [1 emerald for each row]
3. Any suitable answers. **Examples:**
   4,573   4,753   5,347   5,437   [1 emerald each]

Pages 14–15

1. Boxes completed from the left:
   –8   –3   2   7   [1 emerald]
2. a) –3   b) –2   c) –1   d) –4   [1 emerald each]
3. a) –7   b) –2   c) –4   [1 emerald each]
4. Boxes completed from the left:
   –5   10   25   [1 emerald]
5. a) –5°C   b) –2°C   c) –8°C   d) –5°C
          [1 emerald each]
6. –21   [1 emerald]

Page 16

1. a) 58   b) 90   c) 23   d) 79   [1 emerald each]
2. a) XXXI   b) XCV   c) LXVIII   d) XV
          [1 emerald each]
3. a) 187   b) 11   c) 250   [1 emerald each]

Page 19

1. a) 998   b) 793   c) 6,639
   d) 7,576   e) 7,702   f) 9,762   [1 emerald each]
2. a) 210   b) 228   c) 3,212
   d) 2,282   e) 922   f) 2,104   [1 emerald each]

Pages 20–21

1. Lines drawn as follows:
   Sandstone – 125 items   [1 emerald]
   Coal – 30 items   [1 emerald]
   Cobblestone – 175 items   [1 emerald]
   Iron ore – 10 items   [1 emerald]

2 Calculations and estimates matched as follows:
3,512 + 4,894 and 3,500 + 5,000 [1 emerald]
2,967 + 7,370 and 3,000 + 7,400 [1 emerald]
3,459 + 3,202 and 3,500 + 3,200 [1 emerald]
2,732 + 7,998 and 2,700 + 8,000 [1 emerald]

3 a) Estimate: 3,000 + 2,500 = 5,500 [1 emerald]
Actual: 3,022 + 2,536 = 5,558 [1 emerald]
Inverse: 5,558 – 2,536 = 3,022
(or 5,558 – 3,022 = 2,536)

b) Estimate: 8,600 – 4,500 = 4,100 [1 emerald]
Actual: 8,609 – 4,473 = 4,136 [1 emerald]
Inverse: 4,136 + 4,473 = 8,609

4 a) 300 + 300 + 300 = 900, so he has enough granite [1 emerald]

b) 300 + 250 + 250 = 800, so he has enough granite [1 emerald]

c) 450 + 300 + 300 = 1,050, so no he doesn't have enough granite [1 emerald]

Pages 22–23

1 Lower two blocks from the left: 3,525 3,328
Upper two blocks from the left: 6,486 6,853 [1 emerald each]

2 a) 5,885 sand **b)** 8,768 sandstone [1 emerald each]
3 a) 1,960 sand **b)** 6,428 dirt [1 emerald each]
4 a) 2,498 green dye [1 emerald]
b) 6,478 kelp [1 emerald]
c) 1,706 slimeballs [1 emerald]
5 a) 3,972 + 3,210 = 7,182 **b)** 4,058 + 3,603 = 7,661
c) 4,694 – 1,812 = 2,882 **d)** 4,586 – 1,457 = 3,129 [1 emerald each]

Pages 24–25

1

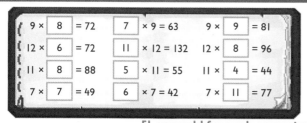

9 × **8** = 72 7 × 9 = 63 9 × **9** = 81
12 × **6** = 72 **11** × 12 = 132 12 × **8** = 96
11 × **8** = 88 5 × **11** = 55 11 × **4** = 44
7 × **7** = 49 6 × 7 = 42 7 × **11** = 77

[1 emerald for each correct row]

2

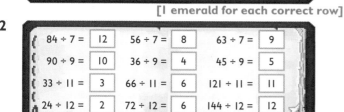

84 ÷ 7 = **12** 56 ÷ 7 = **8** 63 ÷ 7 = **9**
90 ÷ 9 = **10** 36 ÷ 9 = **4** 45 ÷ 9 = **5**
33 ÷ 11 = **3** 66 ÷ 11 = **6** 121 ÷ 11 = **11**
24 ÷ 12 = **2** 72 ÷ 12 = **6** 144 ÷ 12 = **12**

[1 emerald for each correct row]

3 a) 56 [1 emerald]
b) 8 8 × 7 = 56 56 ÷ 7 = 8 [1 emerald]
c) 7 7 × 8 = 56 56 ÷ 8 = 7 [1 emerald]
4 a) 9 × 7 = 63 7 × 9 = 63 [1 emerald]
b) 63 ÷ 9 = 7 63 ÷ 7 = 9 [1 emerald]

Pages 26–27

1 a) From the left: 7 0 78 [1 emerald]
b) From the left: 90 350 0 [1 emerald]

2 a) From the left: 4 6 [1 emerald]
b) From the left: 7 3 [1 emerald]
3 a) 36 **b)** 70 **c)** 120 **d)** 198 [1 emerald each]
4 a) 5 **b)** 120 [1 emerald each]
5 Any suitable answers. **Examples:**
a) (10 × 8) + (4 × 8) = 80 + 32 = 112 [1 emerald]
b) (10 × 5) + (6 × 5) = 50 + 30 = 80 [1 emerald]
6 a) 2 **b)** 5 **c)** 3 **d)** 2 [1 emerald each]

Pages 28–29

1 Circled: 14 is a factor of 28 7 is a factor of 56 [1 emerald each]

2 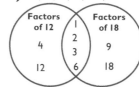 2 × 15

3 × 10   5 × 6

[1 emerald]

30 has eight factors:
1, 2, 3, 5, 6, 10, 15, 30 [1 emerald]

3 a) Circled: 1, 2, 4, 8, 16 [1 emerald]
b) Circled: 2, 17, 34 [1 emerald]
4 a) 1, 3, 5, 15 [1 emerald]
b) 1, 2, 3, 4, 6, 8, 12, 24 [1 emerald]
c) 1, 2, 4, 5, 10, 20 [1 emerald]
5

Factors of 12: 4, 12 | 1, 2, 3, 6 | Factors of 18: 9, 18

[1 emerald for each section]

Pages 30–31

1 a) 195 **b)** 68 **c)** 513 **d)** 325 [1 emerald each]
2 a) 188 **b)** 216 **c)** 216
d) 190 **e)** 287 **f)** 255 [1 emerald each]
3 a) 1,688 **b)** 1,272 **c)** 1,386 **d)** 1,228 [1 emerald each]
4 a) 1,242 **b)** 1,547 **c)** 1,944 **d)** 1,080 [1 emerald each]
5 543 × 7 [1 emerald]
= 3,801 [1 emerald]

Page 32

1 a) 162 iron blocks **b)** 240 emeralds
c) 243 iron blocks **d)** 360 emeralds [1 emerald each]
2 a) 11 chests **b)** 330 emeralds
c) 80 chests **d)** 2,400 emeralds [1 emerald each]
3 a) 168 m **b)** 2,156 m [1 emerald each]

Page 35

1 a) Any 6 segments shaded [1 emerald]
b) Any 6 segments shaded [1 emerald]
c) Any 3 segments shaded [1 emerald]
2 a) $\frac{1}{3} = \frac{3}{9}$ **b)** $\frac{4}{5} = \frac{16}{20}$ **c)** $\frac{7}{8} = \frac{14}{16}$ **d)** $\frac{3}{4} = \frac{12}{16}$ [1 emerald each]
3 $\frac{1}{5} = \frac{2}{10} = \frac{4}{20} = \frac{5}{25} = \frac{20}{100}$ [1 emerald each]

Pages 36–37

1 Any 2 shaded sections + Any 3 shaded sections =
   Any 5 shaded sections   [I emerald]

2 a) $\frac{6}{9}$ (or $\frac{2}{3}$) b) $\frac{7}{12}$ c) $\frac{3}{6}$ (or $\frac{1}{2}$) d) $\frac{9}{10}$
                                                  [I emerald each]

3 a) $\frac{3}{7}$ b) $\frac{3}{9}$ (or $\frac{1}{3}$) c) $\frac{3}{8}$ d) $\frac{2}{7}$ [I emerald each]

4 Any 3 shaded sections + Any 2 shaded sections =
   4 shaded sections and I shaded section in the other
   shape   [I emerald]

5 a) $\frac{9}{8}$ (or $1\frac{1}{8}$)   b) $\frac{14}{9}$ (or $1\frac{5}{9}$)
   c) $\frac{9}{6}$ (or $1\frac{3}{6}$ or $1\frac{1}{2}$) d) $\frac{7}{6}$ (or $1\frac{1}{6}$) [I emerald each]

6 a) $\frac{3}{7}$ b) $\frac{5}{12}$ c) $\frac{6}{8}$ d) $\frac{2}{13}$ [I emerald each]

Pages 38–39

1 a)
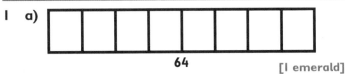
                           64
                                                  [I emerald]
   b) 8   c) 24   d) 56   [I emerald each]

2 a) 4   b) 4   c) 3   d) 12   [I emerald each]

3 a) 45 potatoes   b) 44 carrots
   c) 34 wheat seeds   c) 36 beetroot seeds
                                                  [I emerald each]

4 a) Ticked: $\frac{3}{4}$ of 20 oak planks   [I emerald]
   b) Ticked: $\frac{4}{5}$ of 35 jungle planks   [I emerald]
   c) Ticked: $\frac{3}{4}$ of 60 acacia planks   [I emerald]
   d) Ticked: $\frac{7}{12}$ of 48 spruce planks [I emerald]

5 9 blocks   [I emerald]

6 a) 36   b) 64   [I emerald each]

Pages 40–41

1 a) $\frac{57}{100}$ b) $\frac{19}{100}$ c) $\frac{43}{100}$ [I emerald each]

2 a) Boxes completed from the left:
      $\frac{4}{100}$ $\frac{5}{100}$ $\frac{6}{100}$ $\frac{7}{100}$ $\frac{8}{100}$ $\frac{9}{100}$ [I emerald]
   b) Boxes completed from the left:
      0.73 0.74 0.75 0.76 0.78 0.79 [I emerald]

3 a) 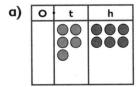   b)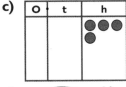

   c)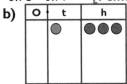
                                                  [I emerald each]

4 a) 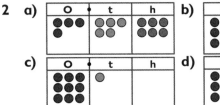 b) c)
                                                  [I emerald each]

5 a) Boxes completed from the left:
      0.54   0.64   [I emerald]
   b) Boxes completed from the left:
      0.86   0.85   [I emerald]

6 a) Boxes completed from the left:
      42.1   42.2   42.3   42.4   [I emerald]
   b) Boxes completed from the left:
      8.99   8.98   8.97   8.96   [I emerald]

Pages 42–43

1 a) 10 0.1   b) 100 0.01   [I emerald each]

2 a) 9.3
   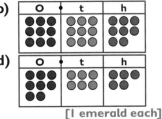
                                                  [I emerald]
   b) 0.19
                                                  [I emerald]
   c) 0.07
                                                  [I emerald]

3 a) 6   0.6   b) 4.3   43
   c) 12 120   d) 4   40   [I emerald each]

4 a) 100 b) 10 c) 100 d) 100   [I emerald each]

5 Shaded: 74 ÷ 10   45 ÷ 100   [I emerald each]

6 Rows completed from the top:
   31 ÷ 100 $\frac{31}{100}$ 0.31   [I emerald]
   68 ÷ 100 Any 68 squares shaded $\frac{68}{100}$ [I emerald]
   4 ÷ 100   Any 4 squares shaded   0.04 [I emerald]

Pages 44–45

1 Largest decimal number with line drawn to it: 7.3
                                                  [I emerald]
   Smallest decimal number circled: 6.8 [I emerald]

2 a) b)
   c) d)
                                                  [I emerald each]
   e) 4.56 < 8.65 < 9.1 < 9.98   [I emerald]

3 a) 4.03, 4.23, 5.24, 9.04, 9.84   [I emerald]
   b) 6.67, 6.53, 5.67, 5.25, 4.52   [I emerald]

4 Maya is wrong.   [I emerald]
   Flower I is the tallest and flower 3 is the shortest.
                                                  [I emerald]

5 a) Any suitable answer. **Example:** 5.53 > 5.07
                                                  [I emerald]
   b) Any suitable answers.
      Examples: 5.37 > 5.05   5.75 > 5.30 [I emerald each]

Pages 46–47

1 Boxes joined as follows:
   $\frac{1}{100}$ to 0.01; $\frac{1}{10}$ to 0.1; $\frac{1}{4}$ to 0.25; $\frac{1}{2}$ to 0.5; $\frac{3}{4}$ to 0.75
                                                  [I emerald each]

2 a) 0.8   b) 0.6   c) 0.08
   d) 0.8 (or 0.80)   e) 0.32   f) 0.94
                                                  [I emerald each]

3 a) $\frac{2}{10}$ 0.2   b) $\frac{4}{10}$ 0.4
   c) $\frac{7}{10}$ 0.7   d) $\frac{9}{10}$ 0.9   [I emerald each]

4 Maya is not correct. 0.3 is $\frac{3}{10}$ as a fraction. [I emerald]

5 a) $\frac{2}{8}$ (or $\frac{1}{4}$)   0.25   [I emerald]
   b) $\frac{4}{8}$ (or $\frac{1}{2}$)   0.5   [I emerald]
   c) $\frac{6}{8}$ (or $\frac{3}{4}$)   0.75   [I emerald]
   d) $\frac{8}{8}$ (or equivalent) I (or 1.0) [I emerald]

6 a) 0.5 b) 0.5 c) 0.25 d) 0.75 [I emerald each]

Pages 48–49

1. a) Boxes completed from left:
   8.1   8.4   8.6   8.9   [1 emerald]
   b) From the left: 8 8 9 9   [1 emerald]
2. Upwards arrows on these chests:
   14.5 kg   16.6 kg   12.8 kg   [1 emerald]
   Downwards arrows on these chests:
   18.3 kg   23.2 kg   27.1 kg   [1 emerald]
3. From the left: 92 46 72 37   [1 emerald for two]
4. From the left: 346 389 268 210   [1 emerald for two]
5. Shaded: 39.8 40.2   [1 emerald each]
6. a) Any three numbers from 59.5 to 59.99… inclusive
   [1 emerald]
   b) Any three numbers from 60.01 to 60.49… inclusive
   [1 emerald]

Page 50

1. 9 km   [1 emerald]
2. a) $\frac{6}{8} = \frac{3}{4} = 0.75$ of the wool is blue.   [1 emerald]
   b) $\frac{2}{8} = \frac{1}{4} = 0.25$ of the wool is red.   [1 emerald]
3. a) 15 kg   b) $\frac{10}{12}$ or $\frac{5}{6}$ c) 6 kg   [1 emerald each]

Page 53

1. Boxes joined as follows:
   Bathtub to 300 l; Sink to 15 l; Jug to 1 l;
   Small pond to 15,000 l   [1 emerald each]
2. a) 2 m   b) 50 cm   [1 emerald each]
3. Any answer from 27–29°C   [1 emerald]

Pages 54–55

1. a) 9 m   b) 200 cm   c) 8 km   d) 7,000 m
   e) 3 l   f) 7,000 ml   g) 3,000 g   h) 4 kg
   [1 emerald each]
2. a) 6 minutes   b) 600 seconds
   c) 36 months   d) 4 weeks   [1 emerald each]
3. a) 1.6 kg   b) 8 m   [1 emerald each]
   c) 320 cm, 450 cm, 5 m, 8 m   [1 emerald]
4. a) >   b) <   c) <   d) =   [1 emerald each]
5. 1,500 ml + 3,000 ml = 4,500 ml; 5 l = 5,000 ml
   [1 emerald]
   5,000 ml – 4,500 ml = 500 ml   [1 emerald]

Pages 56–57

1. 370 m   [1 emerald]
2. Sides measured as: 4 cm, 3 cm, 1.5 cm, 2 cm,
   2.5 cm, 5 cm   [1 emerald]
   Perimeter: 18 cm   [1 emerald]
3. 12 cm   [1 emerald]
4. 60 m   [1 emerald]
5. 20 km   [1 emerald]

Pages 58–59

1. a) A = 33 m²   B = 48 m²   C = 15 m²   D = 76 m²
   [1 emerald each]
   b) C, A, B, D   [1 emerald]
2. a) Any two shapes drawn which each cover 10
   squares of the grid   [1 emerald each]
   b) 10 cm²   [1 emerald]

3. a) 56 m²   [1 emerald]
   b) £224   [1 emerald]
   c) 40 m²   [1 emerald]
   d) £600   [1 emerald]
   e) £824   [1 emerald]
4. a) Any shape drawn of 35–41 grid squares in area
   [1 emerald]
   b) Any shape drawn of 34 grid squares in area and
   symmetrical   [1 emerald]

Pages 60–61

1. a) £2 and 59p   £2.59   [1 emerald]
   b) £2 and 71p   £2.71   [1 emerald]
   c) £5 and 37p   £5.37   [1 emerald]
   d) £4 and 93p   £4.93   [1 emerald]
2. 527p   £5.32   £7.89   £10.87   1184p   [1 emerald]
3. a) >   b) >   c) <   d) =   [1 emerald each]
4. a) £0.25 or 25p b) £2.90   c) £1.05
   [1 emerald each]

Pages 62–63

1. a) Seventeen minutes past two   [1 emerald]
   b) Twenty-two minutes to eleven   [1 emerald]
   c) Twenty to six   [1 emerald]
2. a) 04:35   b) 16:28   c) 00:37
   d) 12:17   e) 03:52   f) 19:26
   [1 emerald each]
3. a) 8:08 pm   b) 3:54 am   c) 11:18 pm
   d) 7:52 am   e) 9:56 am   f) 5:56 pm
   [1 emerald each]
4. a) Thirteen minutes to four   [1 emerald]
   b) 3:47 pm   c) 15:47   [1 emerald each]
5. a) Twenty-six minutes to five   [1 emerald]
   b) 4:34 am   c) 04:34   [1 emerald each]

Page 64

1. a)   b) 11:45 pm   c) 23:45
   [1 emeraldeach]
2. a) >   b) <   c) >   d) <   [1 emerald each]
3. Villager B   [1 emerald]

Page 67

1. a) A: obtuse   B: acute   C: acute
   D: obtuse   E: right angle   [1 emerald]
   b) B, C, E, A, D   [1 emerald]
   c) Any angle drawn that is smaller than angle B.
   [1 emerald]
   d) Any angle drawn that is larger than angle D.
   [1 emerald]
2. a) Circled: angles *a* and *c*   [1 emerald]
   b) Boxes completed from left: < < > [1 emerald each]

Pages 68–69

1. a) Top row, from left: isosceles, equilateral, scalene
   Bottom row, from left: isosceles, scalene, equilateral
   [1 emerald each]

b) Circled: first triangle on top row and second
   triangle on second row   [1 emerald]
c) Ticked: third triangle on top row   [1 emerald]

2 Any suitable right-angled and isosceles triangles.
   Examples:

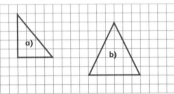

[1 emerald each]

3

Statement	Always	Sometimes	Never
A triangle has one obtuse angle.		✓	
A triangle has at least one acute angle.	✓		
A triangle has three acute angles.		✓	
A scalene triangle has two sides the same length.			✓
All three sides of a triangle are the same length.		✓	

[1 emerald each]

4 **a)**

Side 1	Side 2	Side 3	Triangle?
1	1	6	No
1	2	5	No
1	3	4	No
2	2	4	No
2	3	3	Yes

[1 emerald each]

b) Isosceles (because it has two sides of the same
   length)   [1 emerald]

Pages 70–71

1 Ticked: First and fourth shapes on top row.
   Second, third and fourth shapes on bottom row
                                             [1 emerald each]

2

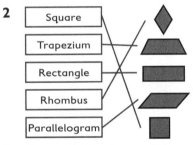

| Square |
| Trapezium |
| Rectangle |
| Rhombus |
| Parallelogram |

[1 emerald each]

3 Any suitable shapes. **Examples:**

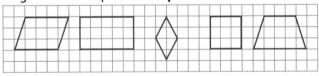

[1 emerald each]

4

	Two pairs of parallel sides	Exactly one pair of parallel sides
Always four equal sides	Square Rhombus	
Not always four equal sides	Rectangle Parallelogram	Trapezium

[1 emerald each]

Pages 72–73

1

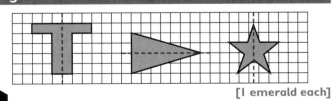

[1 emerald each]

2

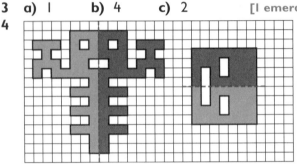

[1 emerald each]

3 **a)** 1   **b)** 4   **c)** 2   [1 emerald each]

4

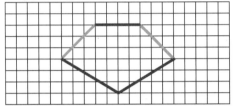

[1 emerald each]

5 Any pentagon with one line of symmetry. **Example:**

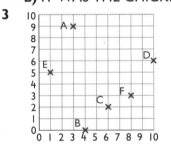

[1 emerald]

Pages 74–75

1 A (1, 2)   B (4, 7)   C (7, 2)   D (7, 10)   E (10, 6)   F (5, 4)
                                             [1 emerald each]

2 **a)** A BAAAD GUY   [1 emerald]
   b) IT WAS THE CHICKEN'S DAY OFF   [1 emerald]

3

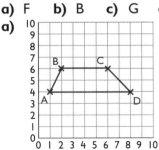

[1 emerald each]

4 **a)** A (8, 4)   **b)** B (6, 2)
   c) C (4, 2)   **d)** D (6, 9)   [1 emerald each
                                             if plotted correctly]

Pages 76–78

1 **a)** C **b)** F **c)** F **d)** B   [1 emerald each]
2 **a)** 2 left and 1 down   **b)** 4 right and 2 up
   c) 2 right and 1 down   **d)** 1 left and 2 up
                                             [1 emerald each]
3 3 right and 2 down   [1 emerald]
4 **a)** F **b)** B **c)** G **d)** F   [1 emerald each]
5 **a)**

[1 emerald]

b) Trapezium   [1 emerald]

6 a)

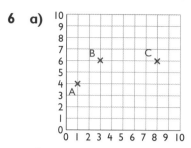

[1 emerald]

b) Point D plotted at and given as (6, 4)   [1 emerald]

7 a)

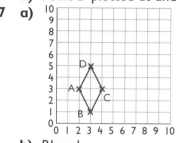

[1 emerald]

b) Rhombus   [1 emerald]

c)

[1 emerald]

d) E (5, 5)   F (6, 3)   G (7, 5)   H (6, 7)   [1 emerald]

Page 81

1 a) 4   **b)** 16   **c)** 6   **d)** 3
   e) Thursday   **f)** 3   **g)** 72   [1 emerald each]

Pages 82–83

1 a) 2   [1 emerald]
   b) Read that the bar reaches halfway between 4 and 6, so it must represent 5   [1 emerald]
2 a) Netherite sword   [1 emerald]
   b) Iron pickaxe   [1 emerald]
   c) 9   [1 emerald]
3 a) Gold shovel   **b)** 3   [1 emerald each]
4

Tool	Number sold
Iron pickaxe	5
Enchanted iron sword	10
Gold shovel	7
Enchanted diamond pickaxe	12
Netherite sword	14

[1 emerald each]

5 48   [1 emerald]

Pages 84–85

1 a) 9 pm   **b)** 2°C   **c)** 3 am   **d)** 7 am
   [1 emerald each]
2 a) 12°C   **b)** 6 hours   [1 emerald each]
   c) i) 5°C   [1 emerald]
      ii) The line is halfway between 4°C and 6°C at the time halfway between 11 pm and 1 am.
   [1 emerald]

3 a)

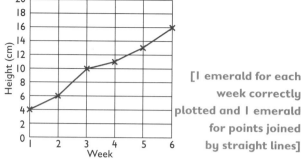

Height of Bamboo Plant

[1 emerald for each week correctly plotted and 1 emerald for points joined by straight lines]

b) Weeks 4 and 5   [1 emerald]

Pages 86–87

1 a) 200   [1 emerald]
   b) Table completed as follows:
      Creepers: 700   Drowned: 1,000   [1 emerald]
   c) The pictogram should be completed with 6 full circles in the Zombies row.   [1 emerald]
2 a) Zombies   **b)** Creepers
   c) 4,600   **d)** 500   [1 emerald each]
3 a) Any suitable axis labels and bar chart title
   [1 emerald]
   b) 200   [1 emerald]
   c) Vertical axis completed in intervals of 200 from 0 to 1,200   [1 emerald]
   d) Horizontal axis completed with Spiders; Zombies; Skeletons; Creepers; Drowned   [1 emerald]
   e) Bars accurately drawn to values of: Spiders 800; Zombies 1,200; Skeletons 900; Creepers 700; Drowned 1,000   [1 emerald]
   Example bar chart:

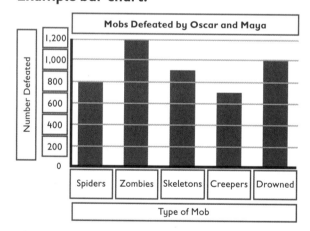

Page 88

1 a) Chicken   [1 emerald]
   b) 12   [1 emerald]
   c) 4   [1 emerald]
   d) 18   [1 emerald]
2 a) 56   [1 emerald]
   b) pigs and cows   [1 emerald]
   c) horses   [1 emerald]
   d) chickens than horses   [1 emerald]

TRADE IN YOUR EMERALDS!

Well done for helping Maya and Oscar succeed in their adventures! Along the way, you earned emeralds for your hard work. This trader is waiting for you to spend your gems with them. Pretend you need to stock up your chests with food. Food is very important for maintaining health. Which items would you buy to make sure you don't go hungry?

If you have enough emeralds, you could buy more than one set of some items.

Write the total number of emeralds you earned in this box:

HMMM?

SHOP INVENTORY

10 RAW RABBITS: 15 EMERALDS

10 RAW PORKCHOPS: 12 EMERALDS

3 BUCKETS OF MILK: 8 EMERALDS

2 BOWLS OF SUSPICIOUS STEW: 6 EMERALDS

1 SPYGLASS: 30 EMERALDS

5 CAKES: 24 EMERALDS

10 NETHER WART: 16 EMERALDS

5 LAVA BUCKETS: 12 EMERALDS

10 GOLDEN CARROTS: 20 EMERALDS

1 PUMPKIN PIE: 5 EMERALDS

1 MOOSHROOM: 25 EMERALDS

10 GLOW BERRIES: 10 EMERALDS

20 CHORUS FRUIT: 20 EMERALDS

10 GOLDEN APPLES: 25 EMERALDS

5 ENCHANTED GOLDEN APPLES: 40 EMERALDS

That's a lot of emeralds. Well done! Remember, just like real money, you don't need to spend it all. Sometimes it's good to save up.